Pete Frolick

SUCCESSFUL DATA PROCESSING SYSTEM ANALYSIS

SUCCESSFUL DATA PROCESSING SYSTEM ANALYSIS

THOMAS R. GILDERSLEEVE

Corporate Systems Development
The Equitable Life Assurance Society of the United States

PRENTICE-HALL, INC.
Englewood Cliffs, New Jersey 07632

Library of Congress Cataloging in Publication Data

Gildersleeve, Thomas Robert.
Successful data processing system analysis.

Includes index.
1. Electronic data processing. 2. System analysis. I. Title.
QA76.G49 658.4'032 78-925
ISBN 0-13-860510-6

Printed in the United States of America

10 9 8 7 6 5 4 3 2 1

Prentice-Hall International, Inc., *London*
Prentice-Hall of Australia Pty. Limited, *Sydney*
Prentice-Hall of Canada, Ltd., *Toronto*
Prentice-Hall of India Private Limited, *New Delhi*
Prentice-Hall of Japan, Inc., *Tokyo*
Prentice-Hall of Southeast Asia Pte. Ltd., *Singapore*
Whitehall Books Limited, *Wellington, New Zealand*

Contents

Part 3
THE FUNCTIONAL SPECIFICATIONS

Part 4
OTHER SKILLS

Preface

The development and installation of data processing systems require many skills. Some of these skills, such as writing programs and designing systems, have a technical orientation. The other skills generally assume some degree of technical competence but are aimed primarily at the human side of systems: the identification of the problem to be solved, the investigation into system feasibility, the functional specification of the system, the training of the personnel who must interface with the system, and the determination of system acceptability. These functions have been classically named system analysis, and they're generally spoken of as being performed by someone acting in the system analyst role.

A person performing in the system analyst role needs skills in two areas:

1. Skills related directly to the procedure he follows to work through to his goals.
2. Skills necessary to perform effectively in carrying out the procedures.

In this book, I ignore the first set of skills completely. A lot of work has been done in this area, it's amply documented in the literature, and we're not going to replow that ground here. The most fruitful approaches in this area appear to be the ones developed under the general rubric of "structured."

Attention will be confined to the second set of skills. Regardless of what system analysis procedure you use, as a system analyst, you must deal with people, you must conduct interviews, you must document procedures, you must make presentations, you must engineer system interfaces for use by humans, and so on. It's with how these things are done that this book is concerned.

Some prior knowledge of system work, either as a programmer, a system designer, a user, or a student, is undoubtedly useful in getting a full appreciation of what is being discussed in this book. It would be fanciful to think otherwise. Nevertheless, I've made every effort to make this book free standing. I've tried to make a minimum amount of assumptions with respect to the prior knowledge and experience of the reader. Thus, this book should be useful to the data processing student with a minimum amount of background in data processing.

On the other hand, I don't mean to imply that this book won't be useful to the experienced system analyst. My experience is just the opposite. I feel that the experienced system analyst will find much useful information in this book.

It may be presumptuous to think that a book could be written which would be useful for the gamut of people from student to experienced system analyst, but that was my goal, and to a significant degree, I believe it was attained. This book is for anyone who aspires to improve his sytem analyst skills.

ACKNOWLEDGEMENT

In three of the chapters of this book I concern myself with topics I've addressed in previous publications. In each case I feel that, in this book, I've organized the material into a more convincing form, significantly expanded the topics considered, and provided materially more detail than before. On this basis, I feel that republication of this material is justified. Nevertheless, in each case, the core of the material remains in the previous publication, and in several instances, I've retained the examples used. Consequently, I want to express my gratitude to the publishers for allowing me to use this material.

1. The chapter on functional specifications derives from Appendix B of *Data Processing Project Management* by T. Gildersleeve (New York: Van Nostrand Reinhold Company, 1974).

2. The chapter on decision tables derives from *Decision Tables and Their Practical Application in Data Processing* by T. Gildersleeve (Englewood Cliffs, N.J.: Prentice-Hall, 1970).
3. The chapter on form design derives from part two of *System Design for Computer Applications* (New York: Wiley, 1963), which I wrote in cooperation with H. N. Laden.

In addition, parts of this book have appeared, in a slightly modified form, as articles in various publications. Once more, my gratitude to the publishers for use of this material.

1. Conducting Better Interviews: *Journal of Systems Management* (2/76).
2. So You'd Like to be More Creative: *Data Management* (10/75).
3. Insight and Creativity: *Datamation* (10/75).

The Equitable Life Assurance system development methodology, of which I give a short description, was developed in cooperation with Hoskyns, which is the owner of the product, System Development Methodology.

Thomas R. Gildersleeve

1

Introduction

The purpose of this book is to help you become a successful data processing system analyst. To achieve this we must do two things.

1. Identify the skills that are generally recognized as those of a successful system analyst whom users trust and want to deal with.
2. Provide a mechanism for acquiring the skills identified—in other words, develop a system analyst training program.

This introductory chapter is devoted to identifying the skills of a successful system analyst. The rest of the book is concerned with the material that makes up the training program.

To begin our task of identifying the skills of the successful system analyst, let us look at the process of developing a data processing system. This process is sometimes referred to as the *system development methodology*.

1.1. THE SYSTEM DEVELOPMENT METHODOLOGY

Most system development personnel agree that there is a system development methodology and that it is made up of phases. Beyond this, agreement begins to evaporate. We have trouble deciding how many phases the methodology has and what their names are. (For some possibilities, see the May 1973 issue of the *EDP Analyser.*) Even if we declare a temporary truce on definitions of phases in a system development methodology, the response to the question of what activities are performed in each phase is bedlam.

In the face of such confusion, it may be instructive to return to basics. To do that, let's try concentrating on that most primordial of applications: payroll. Consider the following statement.

I want a payroll system.

From this we can deduce that the speaker has a payroll or will soon have one. Otherwise, the expressed desire is meaningless.

If the speaker has a payroll, we can assume that it is being met (or the need for it would soon disappear). Consequently we can deduce that he or she already has a payroll system, for this is what is being used to meet the payroll. So by saying, "I want a payroll system," the speaker must mean that he or she wants one of the following.

1. A payroll system that does something different from what the present payroll system is doing.
2. A payroll system that does what the present payroll system is doing, but by some different means.
3. Some combination of (1) and (2).

Thus what is wanted is a change in the system's function, a change in how the function is being carried out, or a change in both.

A change in how a system function is being carried out is a question of what materials are used and of how these materials are combined to carry out the function. The process of putting materials together to make up a system is usually called the *construction* of the system, and the process of deciding what materials will be used in the construction of a system and how these materials are to be combined is usually called *designing* the system. Design is related to the *how* of things; given a function to perform, our choice of the way to do it is a design decision. We will also refer to the person with the "wants" (in the case at hand, the person who wants a payroll system) as the *user.*

So far, we've talked mostly about a user who has a system and wants to change the way in which the function performed by the system is carried out. This, we have said, is a design activity. Now let's hypothesize a user who does not have a system but who wants one. Again, this situation can be visualized in terms of a payroll system. A user who doesn't have a payroll system but expects to need one must be a user who will soon start hiring personnel for the first time. Aside from making the general statement, "I want a payroll system," how can the user specify what he or she wants? Chances are the user has some friends who already have payroll systems, and one way of being more specific is to say, "I want a payroll system like Charlie's." An even better approach may be to look at the payroll systems available from the local service bureaus, select the most appealing one, and enter at least a trial marriage (for better or for worse, your system is my system) with a service bureau.

Suppose the user does take the service bureau approach. This defines the user's payroll system's function (it does what the service bureau's payroll system does) and its design (it carries out its function according to the way the service bureau has constructed it). So in this case the user had a relatively easy job of specifying the payroll system's function and design. This is curious, since our experience tells us that a major problem in system development is getting the user to agree to a functional system specification with which we can work and with which the user will live. Why is the case different here? The answer we favor is the following.

The word "payroll" is a code for a complex of problems, such as determining gross, withholding taxes, and net appropriately; reporting to government agencies when required; reconciling checks; and providing information for management purposes. Over the years this complex of problems has been defined in a fairly precise sense, at least as far as the functions central to the payroll operation are concerned. Consequently when the user says "I want a payroll system," everyone has a relatively clear idea of what problem he or she wants to solve. Thus there is a *universe of discourse,* and the payroll system that the service bureau has designed and constructed to solve the general payroll problem fits the user's requirements rather well. Other areas, such as accounts receivable, accounts payable, and general ledger also have the characteristic of precise problem definition, although even with these areas, and certainly as we move further and further from the classical data processing operations our conviction rapidly fades that we, or anyone else, really grasp the problem with which we're wrestling.

Another reason why the user found it relatively easy to agree to a functional system specification for his payroll when he decided to use the service bureau's payroll system is that, when he agreed to the specification, the system already existed. Consequently he could try it out before buying.

We can see these two factors—clarity of problem definition and existence of a system at specification time—at work in another situation. Suppose the user has an operating payroll system and wants to make a minor improvement in it. This generally presents no major functional specification problem, because the existing system provides a context in which to define the desired new function. The minor nature of the change may also minimize the probability of misunderstanding the problem at which the change is ostensibly addressed; however, it is worth noting that many mindless minor alterations in systems are made on the basis of the argument that the alteration is too minor to justify an inquiry into the problem being attacked. In any case, as the changes become more and more major, the existing system becomes less and less a context in which to define the desired new function, and a clear problem definition becomes more and more essential in order to guard against the catastrophe of a major effort to construct a system that fails to solve the problem. The logical extreme of this extension is the case of a user who wants a totally new system for which no near parallels exist.

What general conclusions can we reach as a result of our investigation of our basic payroll system?

1. For one thing, we've identified at least four different activities in the development of a system.
   a. Problem definition
   b. Specification of system functions
   c. System design
   d. System construction
2. For another, our investigation suggests that, the more the system exists at the time of functional specification, the more likely it is that the user will ultimately be satisfied with the system that is eventually designed, constructed, and delivered.

Let's concentrate for a while on the system development activities we've identified.

1. Problem definition
2. Functional specification
3. Design
4. Construction

These look suspiciously like phase names we've seen in certain system development methodologies. However, we should be cautious about concluding that they *are* system development phases.

If the items listed above are really system development phases, then you must work through them one by one—first you do problem definition, then you do functional specification, then you do design, and finally you do construction. But "Oh, no," someone will argue. "The process of specifying functions may clarify problem definition for you; the design activity may clarify function specification; and so on. You can't compartmentalize these activities into phases—they're interrelated." "Moreover," someone else will leap into the fray, "you can't specify functions without considering how to implement them. Why, you may even have to do some experimental construction before you can ultimately resolve the functional specifications." With all of this we can only agree. So what do we conclude? That a system development methodology is a myth? That there are no such things as system development phases?

Let's try attacking this question from a different perspective. Before we mourn the possible loss of a system development methodology, perhaps we should ask, "Why do we want a system development methodology in the first place?" At least one answer (in our opinion the only justifiable one) is that we want to control system development: We want to keep costs from going through the ceiling, and we want some assurance that eventually we'll get off the system development treadmill and deliver an acceptable, completed product. This brings up the topic of *projects.*

A commonly accepted and apparently useful definition of a project is *an organized effort to reach a predefined goal.* The project is managed by a *project manager,* who has complete responsibility for bringing in the project on time, within cost, and according to specifications. In such a case, the conclusion is inescapable: Phase definition or no, a system development project cannot exist (that is, a project manager cannot reasonably assume his responsibility) until the system has been functionally specified and designed, for until then the project goal has not been defined. Moreover, once the project budget and deadline have been fixed, their integrity depends on stability of the functional and design specifications. Given the fragility of specifications, particularly functional specifications, project management can be a reality only if projects are defined in such a way that project duration is short—say a year or less.

So from the viewpoint of project management, the most we seem to be able to say about a system development methodology is that to permit successful project management:

1. Projects must be restricted to the construction of data processing systems.
2. Functional and design specifications must be frozen before projects begin.

Now let's consider the idea of using system development phases as a vehicle for a *creeping commitment* approach to system development. With such an approach:

1. An initial cost benefit analysis is made.
2. If the cost benefit analysis indicates that the system should be developed, the first phase in system development is undertaken.
3. At the end of the first phase, costs can be projected with greater accuracy. These more accurate costs are used to make a second cost benefit analysis.
4. If this cost benefit analysis indicates that the development should continue, the next phase is undertaken.
5. At the end of this phase, cost projections are again refined, and another cost benefit analysis is made.

Steps 4 and 5 are repeated until either:

1. The cost benefit analysis reaches a state of refinement which indicates that further work should be abandoned, or
2. The last phase in development is completed and a deliverable system results.

The object of this approach is to structure system development procedures in such a way that you avoid overcommitment to work that isn't going to pay off. At the end of each phase you make a feasibility decision on the basis of the cost benefit analysis, and if the information developed in the preceding phase indicates that what originally looked like a good idea has turned sour, you can choose not to throw good money after bad.

There's no doubt that the creeping commitment approach is a sound idea. The question is, Are there natural breaks in system development work that can be treated as phase endings?

Well, so far we've uncovered at least one natural break, when we agreed that successful project management depends on freezing functional and design specifications before projects begin. The freezing of these specifications constitutes the end of a phase, and at this point a cost benefit analysis can be made to determine whether construction of the system should begin.

It also seems clear that an initial cost benefit analysis should be made as soon as is practical after system development begins, to get a first-cut feeling for whether development should continue. But there may not be any natural break in work at which this initial cost benefit analysis is to be made.

Moreover, there may not be any natural breaks in the system development process between the initial cost benefit analysis and the beginning of system construction. Of course, in practice such breaks are defined, and cost benefit analyses are made at these breaks. But we see no compelling evidence that these breaks are natural rather than arbitrary.

Consequently we are willing to sidestep the question of whether problem definition, functional specification, and design are phases of a system development methodology. However, regardless of whether these activities occur one after the other or simultaneously, we maintain that each results in a unique document. Let us say that the activity of defining the problem results in a document called the *problem definition,* the activity of specifying the functions of a system results in a document called the *functional specifications,* and the activity of designing the system results in a document called the *design specifications.* Although all three of these documents can be worked on at the same time (and usually are), the functional specifications cannot be completed until the problem definition is completed, and the design specifications cannot be completed until the functional specifications are completed. We base this claim on the (presumably undisputed) facts that you cannot complete the development of a solution to a problem (the functional specifications) until you've completed the definition of the problem and that you cannot complete a specification of how you intend to do something (the design specifications) until you've completed the specification of what you're going to do (the functional specifications).

Some people maintain that these facts define natural breaks in the system development process.

1. When the problem definition is completed, a natural break occurs. Therefore the work leading up to the completion of this document can be considered a phase in system development.
2. When the functional specifications are completed, another natural break occurs. Therefore the work beginning with the end of the problem definition "phase" and extending through completion of the functional specifications is a phase in system development.

This argument is beguiling. However, we think it falls before the fact that, until the design specifications are done, we never know whether the functional specifications are complete, because in developing the design specifications we may unearth something that makes a change in the functional specifications desirable or necessary. Similarly, until the functional specifications are complete, we never know whether the problem definition is finished, because in developing the functional specifications we may see a change that should be made to the problem definition. So while we can agree that completion of the problem definition and completion of the functional specifications are events that can be used to mark the ends of the phases, we also recognize that these events may occur at times so close to the completion of the design specifications that such phase definition may be impractical. So what conclusions can we draw from our investigations?

1. We have decided that there is at least one genuine system development phase, the *construction phase,* which begins when the functional and design specifications have been frozen.
2. We have also decided that before the construction phase, certain activities called *problem definition, functional specification,* and *design* occur.
3. Finally, we have agreed that the creeping commitment approach to system development is a good idea.
   a. We have agreed that, after the functional and design specifications are frozen, a cost benefit analysis should be conducted to determine whether it is economically feasible to enter the construction phase.
   b. We have agreed that, as soon as is practical, an initial cost benefit analysis should be made to determine whether development of the proposed system is a good idea.
   c. Finally, our discussion seems to indicate that if the system development effort is expected to be significant enough, it might pay to make some feasibility evaluations at other, somewhat arbitrary points in the system development procedure.

For example, there is often a point in system development where, although much design work remains to be done, there is fairly complete agreement between the developers and the user as to what the functional specifications are. In a large-scale development effort, such a point might be a reasonable time to reevaluate feasibility. Similarly, there is frequently a point in system construction where enough of the automated system has been put together to generate some reliable figures on system operating time. A last cost benefit analysis at this point might be advisable.

Because we are placing so much emphasis on freezing the functional and design specifications at the point where the construction phase begins, and because we insist that the design specifications must be frozen before we can consider the functional specifications complete, we should say a few words about what we think goes into the design specifications. As a lead into this discussion of the contents of the design specifications, let's briefly outline the contents of the functional specifications.

In general, the functional specifications will specify:

1. The content of the system output
2. The content of the system records
3. The content of the system input
4. The processing of records and input needed to:
   a. Produce the output
   b. Keep the records up to date

The functional specifications may also include a fairly precise specification of output and input format, which will have quite specific implications for output and input residency.

Given this information from the functional specifications, it remains for the design activity to:

1. Determine the structure and residency of the system records (the data base).
2. Determine the structure of the processing that combines the system's input, data base, and output.

These are the specific data processing system manifestations of the two general design responsibilities.

1. Determine the materials to be used in constructing the system.
2. Determine how these materials are to be combined to perform the system's functions.

(We do not wish to slight many other important functional and design considerations—such as input and output volumes and time schedules, system response and uptime requirements, audit procedures, control techniques, and provisions for security and privacy—but for purposes of this more general discussion we will ignore them.)

One of the raging discussions in the data processing world concerns the extent to which design specifications should specify processing structure. Some people maintain that it should be taken down to the lowest module level, so that the only programming task left is to convert the functional specification of each module into code. Our inclination is in the other direction. We feel that the specification of the processing structure should be taken only to the point where the major functions have been structured. In a batch system, this would be the point where the programs making up the system are identified. Specifications of processing structure should be taken to an equivalent level in a realtime system, although we are not sure how to express this level in terms of realtime systems alone. (Any help here would be appreciated.) We have two reasons for our position.

1. Leaving the specification of program structure to programmers should result in more highly motivated programmers and thus in better constructed programs.
2. The project manager's estimate of system development and system running costs isn't made any more precise by a more detailed set of design specifications. Consequently it's a potential waste of effort to specify a system in more detail when the cost benefit analysis, based on the cost estimates developed at the end of design, may indicate that system development should be abandoned or that the nature of the system should be significantly modified.

Now let's see how our conclusions square with an actual example of a system development methodology. The case we'll examine is the system development methodology of the Equitable Life Assurance Society of the United States. We choose this example because Equitable's methodology has twelve phases, which may well be the largest number of phases in any one methodology. As such, it should incorporate many of the characteristics of the system development methodologies in use today.

1.1.1. An Example

The twelve phases of Equitable's system development methodology are:

1. Project Initiation
2. Initial Survey
3. Feasibility Study
4. Depth Study
5. Business System Design
6. Computer System Design
7. Program Design
8. Programming
9. File Conversion
10. External Procedures
11. Installation
12. Project Review

1.1.1.1. Project Initiation. Equitable says the purpose of the first phase, Project Initiation, is to identify a potential system development that is likely to result in benefits to the user. This usually takes no more than a few hours to complete. Thus this phase involves little more than official designation of a starting point for the project.

1.1.1.2. Initial Survey. Initial Survey recognizes that making judgments about anything depends on having some knowledge about it. The goal of the Initial Survey phase is to obtain preliminary answers to three fundamental questions.

1. What part of the organization will be affected if the proposed changes are made, and what is the current structure of this part of the organization?
2. What are the primary functions of this organization? (The answer to this question is typically phrased in terms of the "big numbers" of the organization: How many people of what different types? How many locations? Volumes of business in dollars or in numbers of transactions, or both? Crucial cutoff dates?)
3. What problems are the changes designed to solve?

If we take a quick look ahead, we see that the next phase (the Feasibility Study) will involve a decision about whether to continue with system development. So in the present phase our example system development methodology is striving to meet one of the requirements of the creeping commitment approach—a preliminary determination, as soon as possible, of the feasibility of developing the system. The goal of the Initial Survey is to gather the minimum of information required for this preliminary determination. Notice also the emphasis on problem definition as a crucial element in the information being collected.

1.1.1.3. Feasibility Study. The Feasibility Study is needed because too much money is often spent on development work before it has been determined that the work is worthwhile. Since it's easy to put off this decision, this phase serves as a reminder that the decision should be made as soon as possible.

1.1.1.4. Depth Study. Depth Study recognizes that if development is to continue, you must know where you're going, and to define that you must know where you are and why you want to leave. During Depth Study the present system is described in detail, and the requirements for a replacement system are specified. Thus in this phase the groundwork is being laid for specifying the functions of the system to be developed.

1.1.1.5. Business System Design. With Business System Design the time has come to decide what the new system is going to be. There will be more opportunities later in development to decide whether you want to continue, but after the Business System Design phase there should be no more questions about where you're going. The object of the phase is to define your goal clearly and completely and to get everyone with an interest in the development to agree to that goal.

During Business System Design, on the basis of the information identified in the Depth Study, a document is prepared that specifies in detail what functions the proposed system will perform. At Equitable this document is called the business system design specifications, but it's equivalent to the document we've been referring to as the functional specifications.

From the user's viewpoint, a major aspect of the proposed system is how he and his staff will interface with it. Consequently the functional specifications must spell out in detail the input and output forms and (if the system has interactive features) the dialogues at the terminals.

1.1.1.6. Computer System Design. Now that the goal has been defined, the next step is to specify how it is to be attained. This is the function of Computer System Design.

1.1.1.7. Construction. At the end of Computer System Design you know where you're going and how you're going to get there. The final step is to go. Program Design, Programming, File Conversion, and External Procedures are all aspects of building and installing the system specified during Business System Design and Computer System Design.

1.1.1.8. Installation. The Installation phase consists of demonstrating that the constructed system does function as specified during Business System Design.

1.1.1.9. Project Review. Strictly speaking, Project Review isn't a phase of system development, which ends with the Installation phase. Nevertheless, taking a look at what you've done to see what you did well and what you could have done better is a crucial step in improving your skills, and this is so seldom done that most phase definitions explicitly state that the step should be taken. Even so, we seldom find the time to perform this valuable function.

1.1.2. Analysis of the Example

Despite the ponderousness of its twelve phases, Equitable's system development methodology is a comparatively good one. Even its ponderousness is mitigated by the fact that the methodology provides for the possibility of tuning it to the system development job at hand. (By *tuning* is meant collapsing the number of phases into a smaller, more appropriate number.) In addition, since in our opinion the Project Initiation and Project Review phases of Equitable's system development methodology really don't represent any part of the system development process, we regard Equitable's methodology as actually having ten phases rather than twelve.

Equitable's methodology avoids two faults sometimes found in other methodologies.

1. The inclusion of a so-called "documentation phase"
2. The inclusion of a so-called "testing phase"

Of course, we have argued that there are really only two natural system development phases (the *specification phase,* which consists of the work done before the freezing of the functional and design specifications, and the *construction phase,* which follows the specification phase) and that all other phases are artificial and thus to some extent arbitrary. Consequently, to dismiss two other phases as distortions of reality may seem inconsistent, but these particular phases can be so harmful that we think they should be drummed out of the system development methodology lexicon.

We reject the concept of a "documentation phase" because we consider system documentation (made up of functional specifications, design specifications, program structure charts, annotated code, operating instructions, user manuals, and system integrity tests) a natural byproduct of the activity of system development. Therefore the inclusion of a particular phase during which documentation is developed guarantees the development of inadequate systems; the development of poor documentation, if any; and demoralized personnel.

That the idea of a "testing phase" is a distortion of reality should be more apparent. Testing permeates the construction phase. As has come to be expressed in the concept of top-down testing, testing is an integral part of computer system construction and can be productively carried back into the design and functional specification activities as well. Testing is also an integral part of the development of the operations and user manuals, and perhaps we could use some phases, similar to top-down testing in computer system construction, to emphasize this fact. Any attempt to segregate testing into a phase separate from the overall system development process leads only to difficulty during development and later, during enhancement, of any existing system.

Equitable's methodology does recognize the one natural break of system development: into a specification phase and a construction phase. In Equitable's methodology this break occurs between the Computer System Design phase and the cluster of phases making up system construction (Program Design, Programming, File Conversion, and External Procedures).

In keeping with the need for creeping commitment, the methodology provides for an early feasibility decision (the Initial Survey and Feasibility Study phases). It recognizes the need for problem definition, functional system specification, and design activities. And by providing a number of (apparently arbitrary) points at which the original feasibility decision can be reevaluated, it carries out the creeping commitment concept in detail.

Now let's take a closer look at the activities that take place during system development.

1.2. SYSTEM DEVELOPMENT ACTIVITIES

1.2.1. The Problem Definition Activity

A clear understanding of the problem that the system is being developed to solve is crucial to the success of a system development effort. Consequently this understanding should be documented. The output of the problem definition activity is the problem definition document.

1.2.2. Initial Activities

The origins of most system development work remain unknown. When do people first become dissatisfied with their present way of doing things? When do they get the first faint glimmer that there must be a better way to do what they're doing or a way to do things that now aren't even being done? When do these promptings and premonitions coalesce into a conviction that action should be taken?

The person first struck by this conviction may be the worker on the job, the worker's boss, a performance review auditor, a consultant, or an analyst or programmer on some other assignment in the area. However, one thing is clear: Whether a data processing system is to be developed is the decision of the user—the line operation that will use the system as a tool in carrying out its mission. Consequently all system development is initiated by a request from the user, even if data processing personnel prompt the user in making it.

One of the first activities in system development is identification of the problem that leads the user to make the request for system development. Related to problem identification is the specification of the objectives that any satisfactory solution to the problem must meet. The objectives must be stated in quantitative terms, because if the system requested is developed, management will want to know whether the system successfully solves the problem. This question requires a quantitative answer.

Also occurring early in system development is the initial functional specification and design work needed as a basis for a preliminary cost benefit analysis of the requested system.

The functional specification work is aimed at getting preliminary answers to two fundamental questions.

1. What is the current structure of the user's organization?
2. What are the primary functions of the organization?

The primary functions are most productively phrased in terms of the "big numbers" of the organization. This information varies from organization to organization but is characteristically concerned with such things as:

1. Number of locations
2. Types of employees
3. Number of employees
4. Number of customers
5. Transaction types
6. Transaction incidence
7. Dollar volume of business

The design activities have two goals.

1. Rough identification of a broad spectrum of alternative design approaches to the problem
2. Selection of one of the alternatives as the most attractive

In doing this design work, the data processing personnel should do whatever is necessary to ensure the technical feasibility of their recommendations and to demonstrate the superiority of the proposed approach to the alternatives. It may even be necessary to write some exploratory code. However, beyond the achievement of these objectives, the design work should be kept to a minimum, since there is no assurance that system development will be carried beyond the initial cost benefit analysis. Any unnecessary development of design detail at this stage is uneconomical.

As a result of this initial functional specification and design work, a system proposal is produced. A system proposal has four parts.

1. *Methodology.* A brief chronology and description of the steps taken in developing the proposal.
2. *Proposed system.* A brief description of the design approach recommended.
3. *Alternatives.* A brief description of the alternative approaches considered. One alternative should be the present method of operation. For each alternative the pros and cons and the reasons for rejection are given. These evaluations are expressed in terms of the organization's "big numbers."
4. *Cost benefit analysis* of the proposed system.

The cost benefit analysis is based on:

1. The cost of operating under the present procedures
2. The benefits of adopting the proposed system
3. The cost of operating the proposed system
4. The cost of developing the proposed system

Acting in a consultant role, the data processing personnel may help the user in developing the first two pieces of information, but the ultimate responsibility for this information lies with the user. The proposed system operating cost will have been developed by the data processing personnel as part of the evaluation of alternative design approaches. Estimating the cost of developing the proposed system is done by the data processing personnel.

After the system proposal has been submitted to the user, a feasibility evaluation is made: Is system development to continue, or is the effort to be abandoned at this time as unfeasible? The evaluation is concerned with the technical, operational, and economic feasibility of the proposed system.

1. Determination of technical feasibility is the responsibility of the data processing personnel. This question should be totally resolved at this point. Proposing a specific design approach means guaranteeing its technical feasibility.

2. Determination of operational feasibility is a user responsibility. At this point the decision can be little more than an educated guess, since the character of the interfaces that the proposed system will present to the user has yet to be considered.
3. Determination of economic feasibility is also a user responsibility. This decision is based on the cost benefit analysis included in the system proposal. Unlike the decision on technical feasibility, which is made at this point and does not arise again, the decision on economic feasibility, if positive, is only an interim decision. The question will arise again during the development of the system.

If the result of the initial feasibility evaluation is positive, then detailed functional specification and design activities are begun.

1.2.3. The Functional Specification Activity

At the beginning of functional specification the emphasis is on a study of the present system. During this study the way the user is presently carrying out the functions that will be replaced by the proposed system is defined in detail, and specification of the requirements of a replacement system begins.

On the basis of the information obtained in the study of the present system, the functional specifications, which specify in detail what functions the proposed system will perform, is developed. In some instances, circumstances allow work on the functional specifications to begin without the intermediate step of present system study, but generally this isn't the case.

From the user's point of view, a major aspect of the proposed system is how he and his people will interface with the system. Consequently, the functional specifications describe in detail the system's input and output forms, and if the system has interactive features, the dialogues at the terminals.

A second major concern of the user is that the proposed system must carry out the procedures for which the user is responsible. Consequently the functional specifications also describe these procedures in detail. However, the procedures are described as the user views them, not as the data processing personnel may view them, and in user terminology, not in data processing terminology.

At the end of system development it must be possible to demonstrate that the system is acceptable—that it correctly performs the functions described in the functional specifications. Consequently the functional specifications also contain the system acceptance criteria—a description of the test procedure the system must correctly complete to demonstrate its acceptability.

The functional specifications document is the output of the functional specification activity.

When the functional specifications are complete, the way the user will interface with the system has been specified. Consequently this is the user's final opportunity to evaluate the system's operational feasibility. Approval of the functional specifications by the user signifies an affirmative answer to the question of operational feasibility. This approval also signifies the user's agreement that the system acceptance test specified in the system acceptance criteria is an adequate demonstration of the system's acceptability.

1.2.4. The Design Activity

The design activity is concerned with specifying the structure of the automated aspects of the data processing system. This structure is documented in the design specifications, which constitute a kind of blueprint for the automated system.

The design specifications provide the basis for a review by all parties concerned with the structure of automated systems. These parties include:

1. A quality assurance function, which is concerned at least with the design's efficiency, security, adherence to data base standards, recovery, and backup
2. The computer center, which is concerned with the design's conformity to computer center operating standards and facilities
3. The maintenance function, which is concerned with the maintainability of the proposed system
4. The auditors, who are concerned with the ability of the proposed system to permit all auditing functions, internal and external

The design specifications document is the output of the design activity.

Once the functional and design specifications are approved, they become the basis for refining the cost estimates for developing and operating the proposed system. These refined estimates are used to prepare a revised cost benefit analysis of the proposed system.

On the basis of this revised cost benefit analysis, the user once more evaluates the system's economic feasibility. This is the user's last opportunity to consider the economic feasibility of the system before system development enters the construction phase, which is generally much more expensive than the specification phase.

1.2.5. The Construction Phase

Construction begins when the functional and design specifications are frozen. Because of the impossibility of freezing these specifications for any significant length of time, the construction phase of system development should not extend for more than a year. Even for sizable systems, if a top-down approach has been taken to functional specification and design, it should be possible to spin off subsystems not requiring more than a year of construction activity. The overall system can then be developed piecemeal, one subsystem at a time.

During construction:

1. The programs making up the automated part of the proposed system are coded and unit tested.
2. The programs are system tested.
3. The files are converted.
   a. A file conversion system is designed.
   b. The programs making up the automated part of the file conversion system are coded and unit tested.
   c. The programs are system tested.
   d. The production operating instructions for running the file conversion system in the computer center are prepared.
   e. The manual procedures related to the file conversion are written.
   f. The personnel who will participate in the file conversion are trained for their role in the process.
   g. The files are converted.
4. The production operating instructions for running the proposed system in the computer center are prepared.
5. The manual procedures related to the proposed system are written.
6. The user personnel who will interface with the proposed system are trained for their role in the production running of the system.
7. The system acceptance test procedure, specified in the system acceptance criteria of the functional specifications, is developed.

8. The acceptability of the constructed system is demonstrated by means of the system acceptance test.
9. The system is installed.

The output of the construction phase is an accepted, operating data processing system.

1.3. ANALYST SKILLS

Data processing system development skills are generally clustered into three groups: programmer skills, designer skills, and analyst skills. This doesn't mean that there are necessarily three types of personnel in a data processing installation, although that is sometimes the case. These skill clusters can be common to a single person; thus there are people who are programmer-designers, or designer-analysts, or even programmer-designer-analysts.

Even in those installations where there are separate jobs (programmer, designer, and analyst), there's some difference of opinion as to which person is expected to perform which function. It seems evident that the functions cluster, but they also appear to overlap. Consequently we can't expect complete agreement as to what functions can be properly classified as analyst functions. In this section we'll describe the functions we believe make up the analyst role.

1.3.1. Project Management

Systems are typically developed by teams of people, the constituency of which changes as the system is developed, calling for different combinations of skills. These teams are managed by a leader, who plans, monitors, and controls the team members' work and coordinates this work with other activities in the organization through communication with management and users. During the construction phase the team becomes a bona fide project, and the team leader a project manager. In some installations the project manager is a designer or a designer-programmer. In others analysts are expected to carry out the project manager role. (For a discussion of project manager responsibilities and how to carry them out, see Gildersleeve, *Data Processing Project Management* [New York: Van Nostrand Reinhold Company, 1974].)

1.3.2. Data Collection

The user is a business expert. The analyst is a generalist on the subject of data processing. To work with the user on the definition of a system, the analyst must learn the user's business. This is done by collecting data about that business, most of which occurs during the early stages of system development, when information is being collected as a basis for the functional specifications. The analyst must be able to extract information from documents, conduct interviews, and understand what is being said. Consequently Part 1 of this book is on data collection.

Part 1 consists of three chapters.

2. Increased Reading Speed and Comprehension
3. Listening
4. Interviewing in Systems Work

1.3.3. Communication

But communication is a two-way street. In addition to collecting data from the user the analyst must communicate to the user the information he has collected, analyzed, and synthesized, so the user can be sure that communication has occurred and that user and analyst are in step on system definition. As the analyst collects data, he must document these facts in a correct and communicative way. He must provide verbal and written feedback to determine that the information collected is, in fact, correct. In particular, he must be able to document previously undocumented procedures in a concise and meaningful way. He must prepare the functional specifications document and communicate its contents in presentations. He generally prepares the user manuals that are used to train user personnel and subsequently become the user's reference on use of the system. If it's the analyst who first sees that there's a better way to do things, he must communicate his ideas to the user: perhaps in documentation, perhaps in formal and informal presentations, perhaps in both. Thus Part 2 of this book is on communications skills.

We have two forms of formal communications: reports and presentations. For a discussion of writing reports and all forms of documents, we refer you to Gildersleeve, *Organizing and Documenting Data Processing Information* (Rochelle Park, N.J.: Hayden, 1977). Consequently Part 2 contains just one chapter.

5. Presentation Techniques

A special type of presentation in which the analyst is involved is the training of user personnel for their role in the production running of the data processing system to be installed. For information on how to prepare such a training program, see Robert F. Mayer and Kenneth M. Beach, Jr., *Developing Vocational Instruction* (Palo Alto: Fearon Publishers, 1967).

1.3.4. The Functional Specifications

The major document prepared by the analyst is the functional specifications. Consequently Part 3 of this book is on the functional specifications.

Part 3 begins with Chapter 6, "The Functional Specifications," which spells out detailed standards for the form and content of functional specifications.

As we have already mentioned, from the user's viewpoint a major aspect of the proposed system is how he and his staff will interface with the system. Consequently the functional specifications describe in detail the system's input and output forms and the dialogues at the terminals, if any. The designer is concerned with forms and terminal dialogue and does receive training on these subjects. However, the emphasis in this training is on how the forms and dialogue relate to the structure of the system. The analyst is concerned with how people relate to the forms and dialogue—the human engineering of the system. Chapters 7 and 8, in Part 3, are concerned with this human engineering of forms and dialogue.

7. Form Design
8. Terminal Dialogue Design

As we've also said, the user is concerned that the proposed system carry out the procedures for which the user is responsible. Consequently the functional specifications also describe these procedures in detail. In the belief that the best way to document these procedures is with decision tables, we devote Chapter 9, the last chapter in Part 3, to decision tables.

1.3.5. Other Analyst Skills

Part 4 addresses some more general analyst skills.

1.3.5.1. Interpersonal Relations. All the analyst's responsibilities, including data collection and communication, require the analyst to work with people continually. Project management requires the ability to get along with people. If the analyst is the first to see that there's a better way of doing things, he or she must be personable enough to convince the user of this. When the user keeps coming up with solutions as system proposals, the analyst must exercise tact in insisting that, instead of implementing the proposed system, time be spent unearthing the underlying problem that is motivating the user to propose solutions.

The analyst is intimately involved in all aspects of the acceptance testing used to demonstrate the acceptability of the newly constructed system. This involvement ranges from the specification of the test (these acceptance test specifications are an essential part of the functional specifications) through the generation of the test data and predetermined results to the coordination of the user, the system development team, and the computer center in test running. The job requires extreme tact and persistence, since the people involved, even if they are philosophically committed to the idea of an acceptance test, may tend to put off the extensive work involved in favor of more immediate demands on their time.

Thus we can see that the ability to work successfully with people has a large bearing on the analyst's success or failure. This subject is addressed in Chapter 10, "Interpersonal Relations," the first chapter of Part 4.

1.3.5.2. Creative Thinking. In specifying the objectives to be met by a satisfactory problem solution, in identifying the primary functions of the user's organization, and in developing the functional specifications, the analyst must think creatively. Perhaps even more importantly, the analyst's creative abilities are called on not only to work on effective solutions to problems but also to be sure that problems are being correctly identified. Consequently Chapter 11, the second chapter in Part 4, is on creative thinking.

1.3.5.3. Cost Benefit Analysis. Finally, the analyst must prepare at least two cost benefit analyses, one when the initial feasibility investigation is being made and another before the construction phase is entered. In addition, if the development effort is divided into additional "phases" to implement a creeping commitment approach, the analyst will be called on to prepare more than this minimum number of cost benefit analyses. The procedure for preparing a cost benefit analysis is described in Chapter 12, "Cost Benefit Analysis," the last chapter in Part 4 and in this book.

1.3.6. System Structure

We began this section on analyst skills by discussing the overlap between system development skill groups. Then we went on to specify the skills that in our opinion make up the analyst role. Several of those who have read this book in manuscript form have expressed surprise that we haven't said more about the process of developing system structure.

Without getting into the debate of whether development of system structure is part of the analyst role, let us simply say that we wouldn't address the subject in this book in any case, because there's already so much good material on the subject available elsewhere. As the best single book to study, we suggest *Structured Design,* by Ed Yourdon and Larry Constantine. Ed is currently publishing this book himself through his firm, Yourdon, Inc., New York, which has the advantage of allowing him to update the book frequently. If Ed follows true to form, this book will eventually be distributed by a major publisher.

Ed also offers a course on "Structured Analysis," which will undoubtedly also lead to a book. Thus, if you want to see how Ed views the analyst-designer role split and to get a good education on the structuring of systems, you might read his book and take his course.

Part 1

DATA COLLECTION

The user is a business expert. The analyst is a generalist on the subject of data processing. To work with the user on the definition of a system, the analyst must learn the user's business. This is done by collecting data about that business, most of which occurs during the early stages of system development, when information is being collected as a basis for the functional specifications. The analyst must be able to extract information from documents, conduct interviews, and understand what is being said. Consequently Part 1 of this book is on data collection.

Part 1 consists of three chapters.

2

Increased Reading Speed and Comprehension

The primary problem with reading is that there's so much to read in so little time. We'd like to find out what we want to know from our reading matter as quickly as possible. Two factors contribute to this speed.

1. Our ability to comprehend what we read
2. The physical speed (in words per minute) at which we read

There's little question that the second of these factors not only contributes to reading speed itself, it also increases comprehension; the faster you read, the more quickly you comprehend. The reason seems to be that the faster you read, the less your mind can wander, the more you concentrate on what you're reading, and thus the more your comprehension is enhanced.

There are five roadblocks to increased reading speed and comprehension.

1. Regression
2. Prolonged fixation
3. Vocalization

4. Pinpoint perception
5. Lack of orientation

Regression is moving your eyes back for another look at something you've already read.

The term *fixation* refers to the fact that when you're actually reading, your eyes are still. You move your eyes when you have read what they're looking at and are ready to focus them on the next thing you want to read. Fixation is necessary for reading, because you can't focus enough for clear vision when your eyes are moving. However, *prolonged fixation,* which is fixing on words longer than is necessary to read them, is a roadblock to faster reading. Instead of just reading the words and moving on, you tend to stare at them.

You break through the regression and prolonged fixation roadblocks by using a *pacer*—a mechanical device to keep your eyes moving as steadily as possible (minimize fixation) and keep them moving forward (prevent regression). The most common and readily available pacer is your hand.

Using a pacer is difficult. It means establishing and maintaining new habits. Education in the use of a pacer won't help you unless you have the dedication to practice both while you're learning the skill and at all times afterward. Consequently we will not say any more about it in this chapter. If you're interested in the subject, you can learn it from almost any good course on speed reading.

Vocalization is the pronunciation of words as they're read. Some vocalizers actually move their lips while they read. Others don't, but subvocalize—they make the sounds in their minds without moving their lips. In any case, vocalization severely restricts the speed at which you can read, since the eye can move and absorb words more quickly than they can be pronounced.

Breaking through the vocalization roadblock is achieved by preventing vocalization from occurring. The most direct way to do this is to chew gum when you read until you've thoroughly broken the vocalization habit.

Pinpoint perception is seeing only what the center of your eye is directly focused on. This may be just a word or perhaps even only a syllable. With pinpoint perception your eye obviously will have to make many fixations to read a line of print. And the more fixations per line, the slower you read.

You break through the pinpoint perception roadblock by recognizing that you do, in fact, see more than what the center of your eye is directly focused on. Physically, you don't have pinpoint perception; you have a span of perception. You're capable of seeing quite a bit to the top, bottom, and each side of the point on which your eye is focused. Pinpoint perception is a psychological restriction, not a physical one. And a severe restriction it is, too.

The fact of the matter is, you don't think a word at a time; you think in phrases—groups of words. To read a word at a time means that you aren't assimilating information in the groupings to which your mind is accustomed. Instead your mind must store the words, one by one, without attempting comprehension, until it has built up what it recognizes as a phrase. Only then can your mind review the phrase and abstract its meaning. A slow and laborious process.

The object is to read a phrase at a time. Then your eye is transmitting information to your mind in its natural groupings, and comprehension occurs immediately. Fix your eyes on the middle of a phrase, use your span of perception to read the whole phrase, and then move your eye to its next point of fixation in the middle of the next phrase. There are two positive benefits of this technique.

1. Fewer fixations per line, and thus greater reading speed
2. Grouping of information in phrases, and thus quicker comprehension

Once you recognize your span of perception, it becomes evident that it's never necessary to fix your eye on either the very beginning or the very end of a line of print. Instead, you should mentally "indent" on both sides of the printed page and move your eyes only between these two margins.

Vocalization and pinpoint perception, and even regression and prolonged fixation, are just bad reading habits and can be broken with practice. Casting off these bad reading habits will greatly increase your reading speed. However, no matter how quickly you read, you'll read without comprehension unless you're oriented to what you're reading. What we mean by *orientation* is as follows.

Given a piece of reading matter—an article or a book—you decide to read it for some reason. The reason is that you have some basis for believing that it contains something you want to know. Formulate the questions you expect the matter to answer before you begin to read. This will help you recognize what the author is saying about what you want to know. Result? Increased comprehension.

If you're familiar with the subject area addressed by the author, you should be able to formulate your questions before starting to read. If you're unfamiliar with the subject but are reading because of the author's reputation or because you want to introduce yourself to a new discipline, skim the matter before reading it. That is, flip through the material—look at the titles, chapter headings, section titles, topic sentences, abstracts, and summaries; decide what you think the author is driving at; on the basis of this review, formulate the questions you want your reading to answer; only then start reading.

If you orient yourself to your reading matter this way, it's frequently possible to "scan" the material and get everything you want out of it. You can skip words, phrases, sentences, paragraphs, sections, and even whole chapters and still obtain answers to all your questions. How much more quickly could comprehension be gained?

EXERCISES

1. What is the relationship between reading speed and comprehension? What is the explanation for this relationship?
2. What is regression?
3. What is fixation? Is it necessary for reading? How can prolonged fixation contribute to decreased reading speed?
4. What is vocalization?
5. How does vocalization decrease reading speed?
6. How do you break the vocalization habit?
7. What is pinpoint perception?
8. How does pinpoint perception decrease reading speed and reduce comprehension?
9. Is pinpoint perception a physical or psychological limitation? Why?
10. What is the alternative to pinpoint perception?
11. Given an article you're going to read, what is the first thing you should do? How should you go about doing it?
12. What is scanning? How can you justify it?

3

Listening

Listening is the ability to understand what another person is saying.

Throughout childhood and adult life, circumstances have encouraged all of us to develop the ability to "tune out" the other person—to avoid listening to what he's saying. Some of us become so adept at this that we don't even physically hear what the other person is saying.

Particularly when we're children, and especially in school, we're subjected to a lot of spoken information that doesn't interest us. All of us (again, particularly when we're children) are exposed to a large amount of verbal criticism, constructive perhaps, but still unpleasant to listen to. And as we grow older, we are practically flooded with an unceasing din of information—commercial messages on radio and television, loudspeaker systems, community meetings, business meetings, education in the form of classes and audio-video presentations, parties—all demanding our unwilling attention. We learn not to listen in order to survive.

There's no reason why your listening shouldn't be selective. But "selective" is the key word. Once you recognize that you have a tendency to tune out the other person, you can learn to resist this general tendency when you *want* to listen.

When do you want to listen? You have goals you want to reach. Some are personal. Some are an integral part of the commercial, political, and social situation in which you find yourself. To reach these goals you need information, some of which is possessed by other people. When you have reason to think someone has information—facts, ideas, attitudes, or feelings—that affect your ability to plan and make decisions for the attainment of any of your goals, then you want to listen. You listen to elicit these facts, ideas, attitudes, and feelings.

The prerequisite to effective listening is the proper listening attitude. This attitude is one of positively pursuing the acquisition of information you don't now have. Here's the way to develop this attitude.

Since you're seeking new information, you must always be prepared for the possibility that what you learn will change your present notion of what things are. The effective listener is one who consistently remains amenable to a change of mind, idea, or approach. You must be willing to inspect objectively each piece of information provided by the other person; don't let any preconceived notions cause you to ignore pertinent information. A good way to keep your own ideas from stifling your receptivity is to make a positive effort to seek out information that conflicts with them.

Treat the other person as your equal. Since he has information you don't have, he's at least your equal in this respect. If you adopt this attitude toward him, you're more likely to understand what he's saying.

Don't let your emotional reaction to words inhibit the reception of information. Words not only have denotation (a reference to the external world) but also have connotation (the emotional freight with which each of us loads words). For example: Is an "incentive" an additional reward for extra effort, or is it exploitation? Is a "negotiation" an attempt to come to an agreement, or is it a method of taking advantage of the other party? When the appeal is to "listen to reason," does this mean you're going to be presented with some new facts, or does it mean that you're going to be subjected to coercion in an attempt to change your mind?

When the other person uses these words, you'll probably start tuning him out if you take "incentive" to mean "exploitation," "negotiation" to mean that the other person is going to try to get the best of you, and "listening to reason" to mean "coercion." Even worse, by word or gesture you may tell the other person that you're no longer listening, which will encourage him to stop talking, and you'll never find out what you need to know.

You can't erase word connotations from your mind, but you can compensate for them.

1. Keep a list of the words that upset you. Draw up an initial list and add to it as experience dictates.
2. Analyze the words on your list. Ask yourself, "Why do I react to this word the way I do?" In many instances, your experience may justify your reaction. We're not asking you to change your reactions, just to recognize why you have them.
3. Then when you react to a word, you can compensate for your reaction. For example, when the other person asks you to "listen to reason," you can still say to yourself, "Oh, oh! Watch out. I'm going to get my arm twisted." But you can then go on to say to yourself, "Well, I'll be on my guard, but let's give this guy the benefit of the doubt. Maybe he's different. If I listen to him, maybe he'll give me some new information."

A common source of emotional freight is stereotypes. Recognizing kinds of stereotypes may help you identify the word connotations that turn you off. Accordingly, we list here some kinds with examples of each.

1. Thing stereotypes
   a. "Those terminals (display units, work organizers, planning tools, etc.) are just toys. They don't help get the work done."
   b. "They'll never improve on the typewriters we have now. All those new features are just gimmicks."
2. Group stereotypes
   a. "I never saw a practical programmer yet."
   b. "Women aren't aggressive enough to be managers."
3. System stereotypes
   a. "Automating an operation never does anything but increase the cost."
   b. "Word processing is a lot of bunk. You'll never compensate for the secretary-boss relationship."

Once you've developed the proper listening attitude, the next step is to get the other person to tell you what you need to know. Here's how.

Always give the other person priority. Whenever the other person starts talking, you stop, no matter how important you think what you have to say is. The other person wouldn't start talking unless he felt he had something to say. You won't find out what you need to know unless you let the other person talk.

Set the other person at ease. The proper listening attitude will do a lot in this regard; someone who feels that you consider him an equal, that you're going to hear him out, and that you aren't going to respond negatively to what he says is encouraged to talk. Also, be pleasant and cheerful; above all, keep smiling; and use appropriate physical contact—the handshake, the pat on the back, etc.

Remove distractions. Common distractions are sounds (tapping, other conversations, and machines) and sights (shuffling papers, flashing lights, and movement). However, setting the other person at ease is more important than removing distractions. If the other person feels more comfortable standing by the printer in the computer room than sitting in a hermetically sealed conference room, then listen to the other person next to the printer.

Use nondirective conversation. If your conversation consists of asking the other person specific questions to which you accept only specific answers, your conversation will be concerned with what you think is important, not what the other person thinks is important. To find out what the other person knows, you must let him talk about what he wants to talk about. As long as the other person talks spontaneously, let him go on. If continuing the conversation requires that you ask questions, ask only the most general questions possible. Leave it up to the other person to decide what's important.

Let the other person know you're listening. Show by your physical posture that you're listening to him. However, fit your posture to the circumstances. If the other person is making a well prepared presentation in which he knows exactly what he wants to say, nothing will inspire him more than rapt attention. However, if the other person is groping for words to express an idea or feeling that he's trying to formulate right in the middle of the conversation, such intense attention may freeze him; a more relaxed, casual attitude on your part is more likely to encourage him to go on. Remember: Give the other person what he needs to keep him at ease.

Withhold judgment at least until the other person has said everything he wants to say. The purpose of listening is to gather information you can use in making a decision. However, don't make the decision in the middle of the conversation. Hear the other person out. If you make your decision before the other person has finished talking, you discourage him from saying everything he thinks is important and as a consequence, you may not learn everything you need to know to make your decision. And remember: Decision making doesn't have to be verbal. A tone or gesture can tell the other person you've made up your mind as easily as can a word. Consequently, maintain your physical listening posture until the other person is finished talking.

Help the other person out. If he has overlooked something, draw it to his attention, so he can take it into consideration in making the judgments he wants to convey to you. However, be as nondirective about this as possible—a good technique is to ask a question rather than to make a statement. Moreover, this is a technique that's applicable only to the later stages of the conversation. Let the other person talk himself out and present the situation as he sees it before you start structuring the conversation with input from your side of the exchange. Finally, remember that each person thinks in his own words, not yours; phrase your questions in his words, so he understands what you're saying.

Take action. You're soliciting information from the other person in order to make decisions. If the other person provides you with information that influences your decision, act accordingly. Nothing will encourage a person to talk to you more than seeing that what he says influences your actions.

Take notes. If, as you should, you want to take action on the new information you've received in a conversation, you must remember the new information. The best way to do this is to take notes during the conversation. Remember that note taking is a two-step process.

1. Taking the notes
2. Rewriting them to make them comprehensible

You should rewrite your notes as soon as possible after the conversation, certainly within twenty-four hours. Then the notes you take during the conversation can be as skeletal as possible, allowing you to minimize writing time and maximize listening time.

Don't let your note taking come as a speech-paralyzing surprise to the other person. Do it habitually, so that people will expect it as a normal part of your listening activity. If the conversation is of the nature of a fact-gathering interview, give your pencil and notepaper prominent display right at the beginning of the interview, so it is clear to the other person that you intend to take notes.

However, listening isn't a passive act of having the proper attitude and encouraging the other person to talk. It's an active search for new information. You must contribute positively to the conversation. What you're looking for in the other person's words are:

1. The principles he endorses
2. The facts he marshalls to support these principles

Here's how to do the most effective job of determining these principles and facts.

Be prepared on the subject to be discussed. Since you expect the other person to provide you with information you don't have, you can never be completely prepared on this subject. On the other hand, since the new information is to bear on a goal you're actively pursuing, there's always some degree to which you can be prepared. Particularly, if the discussion is an interview or a formal meeting that has been scheduled ahead of time, you can prepare explicitly. However, you should always be prepared to some extent on subjects that bear on your goals. Being prepared on the subject makes you more likely to appreciate the true significance of what the other person is saying.

Anticipate the other person. Try to figure out what he is going to say next. If you're right, the point you've anticipated is reinforced. If you're wrong, you can compare how what the other person says differs from what you thought he was going to say, and your appreciation of his point will be heightened by this distinction.

Evaluate what you hear. If the other person is explaining something, ask yourself:

1. Why is he explaining this to me?
2. What is he emphasizing in his explanation?
3. What is he leaving out?

If the other person is making an emotional appeal, ask yourself:

1. Why is he appealing to me?
2. Why is he upset?
3. Is he presenting any facts?

In general, ask yourself, "Do his facts support his principles?" Keep in mind that much of what you hear in a conversation is hearsay, and treat it accordingly. Also remember that when the other person is talking, he's also listening to himself and, frequently, modifying his own thoughts as he talks. A person's first words on a matter are seldom his last.

Pay attention to the context of the other person's remarks. The situation a person is in will influence what he says and how he says it. Take this into consideration when evaluating his remarks. Try to look at the subject under discussion from the other person's point of view. Keep in mind the meaning the other person attaches to the words he uses. If you're not sure, ask.

Remember, what's important is not what the other person says but what he means. True listening is understanding meaning, not words. The other person's tone of voice and gestures may say more than his words. Sometimes the subject the other person avoids is more significant than anything he says. Sometimes what the other person speaks of only with great difficulty is more important than what he can say glibly.

For example, suppose the remark is, "Nobody around here does anything but complain." How would you interpret this remark? Suppose the person speaking is:

1. A plant manager
2. An assembly line worker
3. A black assembly line worker
4. A woman
5. The head of a department store complaint department

Suppose the remark is, "I'll leave it up to you to do this." Does this mean, "I trust you to handle the job"? Or does it mean, "I don't want to have anything to do with it"? Suppose a person holds open a door and says, "After me." What does he mean?

Summarize. Tell the other person what you understood him to say and see whether he agrees with you.

Let's close this chapter with a real problem we all face when listening. The average person speaks at about 125 words per minute. You can think much faster than that. The question then is, What do you do with all that spare thinking time while you're listening? One thing you can do is occupy the unused part of your mind with your private thoughts. The trouble with this solution is that you'll slowly drift away from the conversation to the point where you're not giving it the attention the subject deserves. As a result, you won't hear the information for which you're looking. What's more, your inattention will discourage the other person from talking, and he may stop before he says the very thing you need to know.

The productive approach is to pursue your listening role actively. If you do, then instead of falling into the trap of becoming inadvertently inattentive, you'll be figuratively moving ahead of, behind, and to the sides of the topic under discussion as the other person talks. You'll be anticipating what he's going to say, reviewing what he's said, and analyzing the entire conversation to see how it fits together.

EXERCISES

1. When do you want to listen to someone?
2. How can you develop the proper listening attitude?
3. How can you get the other person to tell you what you need to know?
4. How can you actively contribute to a conversation while you're listening?
5. Describe how you go about determining:
   (a) The principles the other person endorses
   (b) The facts he uses to support these principles
6. You can think much faster than the other person can talk. While you're listening, what should you do with your spare thinking time?

Interviewing in Systems Work

Interviewing is a form of conversation. People engage in conversation for a variety of reasons.

1. They enjoy it.
2. One person wishes to impart information to another.
3. One person wishes to collect information from another.
4. One person wishes to change another person's behavior or attitude.

Interviewing is a form of the third type of conversation. The person who wishes to collect information is called the *interviewer;* the person from whom the information is to be collected the *respondent.*

There's nothing wrong with the respondent and interviewer enjoying the conversation that goes on during an interview. In fact, the interviewer will be more successful in obtaining the information he's seeking if both he and, particularly, the respondent enjoy the interview. Nevertheless, an interviewer who uses the interview as an opportunity to socialize with the respondent is not furthering the goal of the interview, which is to collect information.

It's also true that the chance for a successful interview is enhanced if the respondent wants to communicate the information that the interviewer is seeking. Therefore one of the interviewer's goals is to engender such an attitude in the respondent. However, in preparing for the interview the interviewer must assume that the interview will occur only because he's going to initiate it. The interviewer can't assume that the respondent will voluntarily supply the information for which he's looking.

So while the interviewer should encourage the respondent both:

1. To enjoy the interview and
2. To want to provide the information for which the interviewer is looking

the interviewer must still take the initiative in getting information from the respondent. However, the interviewer should never consider the interview an opportunity to try to change the respondent's behavior or attitudes. The situation in which one person converses with another to elicit information *and* change behavior is called *counseling.* The approach a counselor takes to a counseling session is, "Let's see if we can find out what's wrong, and then let's see what we can do to correct the situation." Interviewing is a technique used to determine facts. Frequently, the facts to be determined are, "What went wrong?" However, unlike a counseling session, an interview should never involve any implication that the respondent may in any way be part of the problem (even if he is) or that the goal of the interview is to "improve" the respondent's behavior in some way.

Since the interviewer must initiate the interview and since communication is successful only if both parties participate, the first problem facing the interviewer is one of motivation: How can he get the respondent to participate willingly in the interview, let alone enjoy it or be the initiator in providing the information?

At the outset of the interview, the interviewer typically doesn't have to provide this motivation himself. Frequently the respondent either has been instructed by his superior to participate in the interview or has volunteered to participate. Even if such isn't the case, in our society the interviewer doesn't immediately face the problem of providing motivation; almost anyone who is faced with someone trying to attract his attention is schooled to extend the common courtesy of listening while the intruder states his case.

However, whatever initially motivates the respondent to give his attention to the interviewer, the interviewer can be sure this motivation won't last long. Therefore the interviewer must use this opportunity to engender some other, longer-lasting type of motivation in the respondent.

What form might this motivation take? To answer this question, let's structure the situation in more detail. We're concerned with a situation in which a system analyst is conducting an interview to collect information that will be used to improve his company's systems and procedures. Some reasons why a respondent would be motivated to participate constructively in such an interview might be as follows.

1. He likes to talk.
2. He likes to demonstrate his knowledge of the subject in which the interviewer is interested.
3. He wants to ingratiate himself with his boss.
4. He wants to contribute to the improvement of company procedures, which is the long-range goal of the interview.
5. He feels it's part of his job.

There are probably as many motives for participating in an interview as there are respondents, but these are some of the most usual ones. It's interesting to note that in each case but the first, before the respondent can satisfy his motive (that is, before he can demonstrate his knowledge in the pertinent subject area, ingratiate himself with his boss, do his job by providing the information desired, or contribute to the improvement of company procedures), he must know:

1. What the procedures to be improved are
2. What bearing the information he is to supply has on improving these procedures
3. How he is to supply the information

In the case of the respondent who just likes to talk, there's no problem in getting him to participate in the interview. However, even in this case, if he's to participate constructively, the topics of the discussion must be channeled. He must be made aware of the procedures to be improved, the bearing of the desired information on the improvement to be made, and how he's to supply the information.

Thus it becomes clear that the most constructive way the interviewer can use the attention the respondent initially gives him is to *orient* the respondent to the interview, that is, to tell the respondent:

1. What procedures the interviewer is working to improve
2. How the respondent can help in making these improvements by supplying certain information
3. What information, in general, the interviewer needs
4. How the respondent is to provide this information

This kind of orientation provides the respondent with a sound reason for making the desired information available.

However, even if he has been properly oriented, the respondent will communicate only if he feels that he'll be constructively heard. Therefore by his words and actions the interviewer must convince the respondent that:

1. The two of them have some overlap of knowledge in the area under discussion.
2. The interviewer is capable of constructively using the information supplied.
3. The interviewer respects the respondent's ability to provide the information needed.
4. The respondent is free to express himself without fear of being judged by the interviewer.

This attitude on the part of the respondent isn't one the interviewer can create full-blown at the beginning of the interview. It must be cultivated throughout the interview. As the interview proceeds, the respondent's confidence in the interviewer should grow.

Once the interviewer provides the respondent with adequate motivation to participate constructively in the interview, it remains for the interviewer to be certain he obtains the information he needs.

Since the interviewer's goal is to obtain information from the respondent, it follows that the interviewer never knows exactly what information the respondent has. Since it's the interviewer who directs the interview, it's always possible that the interviewer will fail to obtain some information the respondent has simply by failing to give the respondent the opportunity to present the information. As a consequence, it's a general rule in the conduct of interviews that after the orientation, the interviewer should:

1. Start the questioning process with the most general of questions in the subject area.
2. Follow up on the topics the respondent raises.
3. After having exhausted the topics raised by the respondent, start investigating areas not already touched on in the interview by asking more specific questions.

Thus the general progress of the interview is from the general to the specific.

To be sure he covers all the areas he should, the interviewer should prepare himself before the interview. The best way to make this preparation is to develop an *interview guide,* a written set of questions that outlines the areas the interviewer thinks should be covered in the interview.

The most common difficulty with which the interviewer must contend is the incomplete answer.

Sometimes the respondent answers some question other than the one asked. In this case the appropriate interviewer action is to clarify the situation politely and repeat the question.

If the respondent's reply is too brief or too general, one way of getting him to expand on his remarks is to summarize what the respondent has said. For example:

> Interviewer (I): Do you find the information you're getting on our sales activity satisfactory?
>
> Respondent (R): Not by a long shot.
>
> I: Then you're dissatisfied.
>
> R: I'll say I am. As it is now, it's practically useless.
>
> I: You can't use the information the system is giving you.
>
> R: Not the way it comes to us. First we have to summarize it by territory and salesman. Then we can begin to see where we stand.
>
> I: So the system would be more useful to you if the sales activity were summarized by territory and salesman.

Here, simply by summarizing what the respondent has said, the interviewer gets him to be more specific.

Another technique for getting a respondent to expand on his remarks is the "nondirective probe." Three types of nondirective probes are as follows.

1. A brief assertion of understanding or interest, such as, "I see," or "Um-hm."
2. A pause in which the interviewer says nothing at all. Such a pause often encourages the respondent to expand on his previous comment.
3. A neutral phrase, such as:
   How do you mean?
   I'd like to know more about your thinking on that.
   What do you have in mind there?
   I'm not sure I understand what you have in mind.
   Why do you think that's so?
   Why do you feel that way?
   What do you think causes that?
   Do you have any other reasons for feeling as you do?
   Anything else? [1]

Since the interviewer is gathering information from the respondent, the respondent may give the interviewer an answer he doesn't understand. Under these circumstances, the interviewer must ask the respondent to clarify his remarks. For example, if the respondent uses a term unfamiliar to the interviewer, he must ask the respondent to define the term.

Sometimes the respondent just doesn't understand the interviewer's question. In such a situation the interviewer must reword the question to clarify what he's asking.

Sometimes the respondent is confused as to what his position is on a given topic. Then the interviewer must try to help the respondent clarify his thinking.

Kahn and Cannell [2] give three scales for rating the effectiveness of a probe in eliciting further information from a respondent.

1. *The acceptance scale.* Does the probe contribute to the interviewer-respondent relationship, or does it tend to reject the respondent or his response?
2. *The validity scale.* Does it leave the respondent free to expand on his comments as he sees fit, or does it tend to force him to make a specific response that may not be representative of his true feelings?
3. *The relevance scale.* Does the probe encourage the respondent to provide the information that the interviewer is seeking, or does it tend to lead the respondent away into irrelevant material?

DeMasi [3] provides a table of difficulties with which an interviewer may be faced in an interview and the action he can take to overcome the difficulty (see Table 4.1).

Some questions carry more of a threat to a respondent than others. For example, asking a respondent to evaluate his work performance may appear threatening to him. The danger of a threatening question is not only that the respondent may evade the question but also that he may become uncooperative for the remainder of the interview. Consequently, if threatening questions must be asked, they should be the last ones in the interview, for two reasons.

1. If the respondent does react negatively to the question, the interviewer will have at least gotten cooperation from the respondent in his answers to the preceding questions.
2. The course of the interview up to the potentially threatening question will have given the interviewer every possible chance to develop the respondent's confidence in the interviewer. As

Table 4.1

Respondent Behavior	*Interviewer Action*
1. Appears to guess at answers rather than admit ignorance.	1. After the interview, cross-check answers that are suspect.
2. Attempts to tell the interviewer what he presumably wants to hear instead of the correct facts.	2. Avoid putting questions in a form that implies the answers. Cross-check answers that are suspect.
3. Gives the interviewer a great deal of irrelevent information or tells stories.	3. In friendly but persistent fashion, bring the discussion back into the desired channel.
4. Stops talking if the interviewer begins to take notes.	4. Put the notebook away and confine questions to those which are most important. If necessary, come back later for details.
5. Attempts to rush through the interview.	5. Suggest coming back later.
6. Expresses satisfaction with the way thing are done now and wants no change.	6. Encourage him to elaborate on present situation and its virtues. Take careful notes and ask questions about details.
7. Shows obvious resentment of the interviewer, answers questions guardedly, or appears to be withholding data.	7. Try to get him talking about something that interests him.
8. Sabotages the interview by noncooperation. In effect, refuses to give information.	8. Ask him, "If I get this information from someone else, would you mind checking it for me?" Then proceed on that plan.
9. Gripes about his job, his pay, his associates, his supervisors, the unfair treatment he receives.	9. Listen sympathetically and note anything that might be a real clue. Don't interrupt until he has poured out his gripes. Then make friendly but noncommittal statements, such as "You sure have plenty of troubles. Perhaps the study can help with some of them." This approach should bridge the gap to asking about the desired facts. Later, make enough of a check on his gripes to determine whether there is any foundation for them. In this way, you neither pass over a good lead nor leave yourself open to being unduly influenced by groundless talk or personal prejudice.
10. Acts as eager beaver, is enthusiastic about new ideas, gadgets, techniques.	10. Listen for desired facts and valuable leads. Don't become emotionally involved or enlist in his campaign.

a result, a question that at the beginning of the interview might have seemed threatening to the respondent may no longer appear so.

When terminating an interview, the interviewer must do three things.

1. Give the respondent one last opportunity to mention anything he thinks is significant that hasn't yet been discussed.
2. Get the respondent's permission to return with any questions of which the interviewer may subsequently think.
3. Get the respondent's commitment to review the interviewer's interview report.

Here's an example of a well conducted interview. The situation is as follows.

Dick Tater, Executive Vice President of Hurd Instruments, Inc. (more commonly known as Hurd Inst., Inc.), recently returned from a cram update education program on the latest developments in management techniques conducted by the local graduate business school. At this session the Marketing Vice-President of another concern made a presentation on how his company was using automated data processing to increase their marketing productivity. As a result Dick asked Hugh D. Knee, his Director of Data Processing, and Prometheus Motor, his Marketing Vice-President, to investigate cooperatively the feasibility of a similar approach at Hurd Instruments. Hugh delegated the responsibility for Data Processing's end of the investigation to Sam Spade, one of his analysts. To give Sam a basic understanding of what marketing productivity is and how it might be improved, Hugh made an appointment for Sam to speak with "Pro" Motor. We overtake Sam as he enters Pro's office to keep his appointment.

Here's a transcript of the interview that followed. In this transcript comments on the techniques used by the interviewer are in brackets.

I: Good morning, Mr. Motor. I'm Sam Spade, from Data Processing.

R: Oh, yes. Hugh said you'd be over. I'm pleased to meet you.

I: The pleasure's mine, Mr. Motor.

R: Just call me "Pro," son—all the boys do. Have a seat.

I: Thank you.

R: Now, what can I do for you?

I: Well, sir, you know that Mr. Tater has asked us to investigate the feasibility of using automated data processing techniques to increase our marketing productivity. Hugh has given me the responsibility for conducting this investigation from Data Processing's point of view, and to be perfectly frank with you, while I flatter myself that I know something about data processing, I know very little about marketing. Hugh and I thought that, if I could just sit down with you for a while, you could give me a basic understanding of what marketing productivity is and then perhaps I could begin to see how data processing might be able to help in improving it.

R: Well, that sounds like a reasonable approach. Any particular place you'd like to start?

I: No, sir. If you don't mind, I'd rather leave that up to you.

[Sam has wisely allowed Pro to pick the ground on which he wants to start.]

R: Good enough. Perhaps we should start by defining marketing. Hurd Instruments' goal is to satisfy the public's needs at a profit. In Marketing, we're concerned with those activities that result in the flow of products from us to the organizations and individuals who can use our products. Here are some examples of marketing decisions: Deciding who the prospects for a product are—how many there are, where they're located, and what characteristics they have in common; the methods we're going to use to deliver the product to our customers; the way we're going to package the product—what options we're going to offer, whether we're going to combine it with other products, what the minimum order is going to be; how we're going to price the product; what kind of selling techniques we're going to use—are we going to use salesmen or agents, and what kind of commission structure are we going to support? What advertising and promotion techniques are we going to employ?

Maybe the first point I should make about marketing is that it's important. The saying "Build a better mousetrap and the world will beat a path to your door" just isn't true. You won't sell many mousetraps if your price is too high or if you sell them only by the gross.

Neither will you sell many if you can't get people interested in eliminating mice, if they can solve their problem without traveling to your door, or if they don't know where your door is.

Second, marketing isn't easy. Marketing problems may appear simple, and their solutions may look like a matter of common sense. But this is deceptive. For example, something as elementary as an advertisement represents decisions about who the message should appeal to, what medium is most likely to carry the message to the audience, what should be contained in the copy, and how much should be spent in this particular aspect of the overall marketing campaign—which implies a whole host of other questions concerning what the overall marketing strategy is and how this advertisement fits into it. However, marketing is certainly easy in one respect: It's easy to make bad and costly decisions.

Well, perhaps that's enough by way of introduction. Where do we go from here?

[This is a crucial point in the interview. Pro is being quite expansive, and given the opportunity, there's no telling in which direction he might run. Notice that, despite the fact that Sam used the term twice in his introduction, Pro has not yet addressed himself to the topic of marketing productivity. After having let Pro open up, Sam is now going to lead him to the specific topic in which he's interested.]

I: That's certainly a fine start, Pro. In a couple of minutes you've given me a good general feel for the wide range of problems with which you people in Marketing deal. I'm particularly struck by the point you make about how easy it is to make a bad marketing decision. It seems to me that that's where the concept of marketing productivity comes in. You can't really avoid making marketing decisions, and anytime you can improve the decisions you'd otherwise make, you've improved marketing productivity. Is that right?

R: That's the idea, Sam.

[Here a pause ensues. Finally, Sam goes on.]

I: OK. I'm sure glad I've got the general idea. But the whole thing is still pretty hazy to me. Would you mind telling me in your words what marketing productivity is?

[Whew! Sam has successfully skirted another problem. In leading Pro to the topic of marketing productivity, Sam asked Pro a yes-no question, and Pro gave no more than the yes answer. However, Sam has now recovered by asking Pro to expand on his answer.]

R: Be glad to. It's fairly clear that it costs money to conduct a marketing effort. As you point out, no matter how ineffectively we do it, we'll be spending time determining our prices and packaging, and we'll be spending money on our product distribution, selling, and advertising efforts. So one factor in the marketing productivity equation is marketing cost.

The other factor is profit. Remember, we said marketing is the satisfaction of needs *at a profit.*

Marketing productivity is the ratio of profit to marketing costs. The bigger the ratio, the better the productivity.

I: Let's see if I've got this right. Theoretically, there are three ways we can improve marketing productivity.

1. We can increase marketing costs as long as this results in even larger increases in profits.
2. We can reduce marketing costs as long as this doesn't result in a proportionate decrease in profits.
3. We can hold marketing costs constant, but redistribute our marketing effort so as to increase profits.

R: Hum . . . yeah . . . I guess you could say that.

I: You don't sound completely convinced.

[With this probe, Sam investigates Pro's hesitation. By doing so, he'll uncover some qualifications of which he might not otherwise have become aware.]

R: Well, what you're saying follows from what I told you. But in trying to spell out what marketing productivity means, I'm afraid I oversimplified the situation. There are several qualifications I'd like to make.

First, what you say would be true if we marketed only one product in only one market. However, we market many products, and many of these we market in many different market areas. So what you say holds true for one product in one market, but in general, an increase in Hurd's overall marketing productivity would probably result from all three strategies operating in various product markets.

Another qualification in this: It may well be demonstrable that a change in marketing plans for a given product in a given market will improve profit for that particular product-market combination, but the change may still be inadvisable because it will change the product mix so as to, for example, increase production costs in other product areas and result in an overall decrease in company profitability.

And as a last qualification, I'd mention that even though a reduction in marketing cost may improve marketing productivity by resulting in a less than proportionate reduction in profits, management may, for financial or long-range planning reasons, still not want to make the cost reduction because of the resulting reduction in total revenues.

I: Those are certainly good points and I'm glad you made them. They help put the concept of marketing productivity into perspective. Now, how can data processing be of help in improving marketing productivity?

[Sam has just delivered the other half of his one-two punch. Having got Pro solidly on the subject of marketing productivity, he now introduces the topic of data processing as related to marketing productivity. If Pro will follow Sam's lead here, Sam will have accomplished the first goal of interviewing: He will have oriented the respondent to the subject to be discussed in the interview.]

R: I'm not sure I'm qualified to answer that question. Maybe you can help me here. If I remember correctly what I was told in that data processing orientation workshop Dick made us all go to last fall—when it's structured right, the normal processing of the activities that go on at Hurd create a base of data that reflects our experience over time, and we can organize and summarize this data in various ways in the attempt to uncover information we can use to improve our performance. Is that the idea?

I: It sounds to me like you've got the message.

R: OK. First of all, I'm sure our normal operations in Marketing don't generate all the kinds of information we need to improve marketing productivity, but they should, and for the moment, let's assume they do. Now let's perform a little experiment. Let's suppose we increase the marketing costs for any single product over time from one period to the next. In theory, at first, both sales volume and contributions to net profit should go up with the increasing marketing effort. But eventually two things are going to happen.

1. Sooner or later, sales volume will tend to increase at a decreasing rate—flattening out and approaching (but never reaching) 100 percent of market share at its upper limit.
2. Sooner or later, contributions to net profit will reach a maximum, then decrease, and then become negative.

If we had the marketing data we need, then as we collect data during the course of our experiment, the data we collect should trace the patterns the theory describes, and as a result, we should be able to tell by inspecting the resulting data what the optimum marketing investment is—where an increase of marketing cost no longer results in a significant increase of market share, and where an increase in marketing costs no longer results in an increased contribution to net profit, but instead begins a decline in that contribution.

But we could undoubtedly improve our marketing effectiveness even without conducting any experiments. If we just had sufficient cost and profit information on all our products in all markets, we might be able to do all sorts of interesting things. For example, we might find that:

1. Contrary to the unquestioning adherence to the idea that "Customers are the lifeblood of business," the business we do with a significant number of customers may be distinctly unprofitable.
2. The most profitable channel of product distribution may be neither the one that produces the most volume nor the one that has the lowest unit cost.
3. The breakeven point between an order that's unprofitable and one that's profitable may lie at a much different order level than we ever expected.
4. There may be wide variations in the ratio of sales to regional potentials as between different territories of our market.
5. The advantage of more complete utilization of facilities and personnel may be more than offset by the difficulties of mounting a marketing campaign for a multitude of dissimilar products.
6. Simply repackaging a product may have a marked impact on its profitability.

I: Wow! The potentialities boggle the mind. Seriously, Pro, I'm afraid you left me in the dust there somewhere. I'm all confused as to products, markets, customers, distribution channels, orders, territories, product mix, and packaging. Can you back up and try to straighten me out? I'll try to pay better attention.

[Everything Pro has said about marketing productivity is true, but he's addressed himself to the big picture, and he's tended to lapse into marketing jargon. Sam has wisely asked Pro to clarify his remarks.]

R: Well, I will admit I got a little carried away there. Let's try to take it a piece at a time.

Suppose we have only one product, that we sell it only on a unit basis, that we sell it only to gas stations, and that the salesmen who have the responsibility for selling our product in the first place are also responsible for stocking it in the station after it's been sold. We would then have only one package, one customer type, one distribution channel, and one promotion technique. We would have no product mix. There would be only two variables: order size and geographic area. If we could then collect marketing cost and profit data by order size and by geographic area, we could analyze this data. Perhaps we'd discover that:

1. We make much more profit in some geographic areas than others, and/or
2. Orders below a certain size are unprofitable.

As a result, we might decide not to cover certain geographic areas and also to refuse orders below a certain size—assuming of course, that such a policy wouldn't have an adverse effect on our other orders.

Of course, even this example isn't so simple, since we would have to determine what makes up a significant division of the country into geographic areas. We might have to collect data on the basis of many different divisions and neutralize the impact of the varying abilities of our salesmen before we could decide which divisions are significant from a marketing productivity point of view.

Now let's suppose that, in addition to selling our product to gas stations, we also decide to sell it to supermarkets. We've now introduced another variable—customer type—and to determine its effect on marketing productivity we're going to have to be able to break down our costs and profits by customer type as well as by geographic area and order size. If we decide to deliver our product by mail too, we've introduced another variable. A decision to institute an advertising campaign introduces another variable. And so on.

Now, if we introduce more than one product, we need to collect cost and profit information for each product on each market variable. We might decide to offer combinations of products in packages, which introduces yet another variable. And so it goes.

Any clearer?

I: Yeah. Now I really think I've got it. For each product there are a number of market variables that have an influence on sales volume. Some examples of market variables would be customer type, distribution channel, promotional techniques, order size, geographic area, and packaging. As a first step in measuring marketing productivity it's necessary to collect marketing costs and profits for each product on each one of these market variables. Is that what we're saying?

[To see if he really has the total picture, Sam has now summarized what Pro has told him. And it's a good thing that Sam does make this summary, because as we shall see, it causes Pro to introduce some significant qualifications.]

R: It sure is. But just in case you think the situation is getting too simple and straightforward, let me throw a couple more curves at you.

First of all, getting a handle on contribution to net profit by product and market area is a very complex calculation. It involves not only the revenue and marketing cost figures we've been talking about, but also all of Hurd's production costs for the product plus the fixed costs of staying in business.

I: I can see that even for a computer, that would be a pretty complex calculation.

[Sam has probed to see if Pro has anything more to say on the subject.]

R: Fortunately, there's a relatively easy way to avoid this complication. Instead of insisting on a contribution-to-net-profit figure, we settle for an indication of the figure. The indicator typically used is the difference between revenue and marketing cost.

I: So we're back to the basic data we were originally concerned with.

R: Right.

I: OK. So much for contribution to net profit. Now, you said you were going to throw me a couple of curves. What's the other one?

[Sam has followed up on a topic Pro had previously mentioned.]

R: The other has to do with those marketing costs we've been talking about so glibly up to now. You see, some marketing costs can be directly related to the marketing of a given product in a given market area. But unfortunately, other marketing costs cannot. We call this latter type of costs "indirect costs," and examples of indirect costs include such things as warehousing, shipping, and order processing.

I: Hum . . . and how do we work these indirect costs into our marketing cost determination?

[Sam has recognized that Pro's presentation on determining marketing costs is incomplete, so he has urged Pro to go on.]

R: Well, as indicated by the examples I just gave you, indirect costs can typically be collected and summarized by function. Some other examples of functional areas that contribute to overall marketing costs are accounts receivable, product planning, packaging, pricing, advertising, and market research. These indirect costs must be allocated on the basis of a judgment as to what percentage from each functional area can be charged to the marketing effort for each product in each market area.

I: So from a procedural point of view, what we have to do is accumulate indirect costs by functional area and allocate these indirect costs by product and market area; then for a given product in a given market area, we can sum both the direct and indirect costs to arrive at a total marketing cost, and when we subtract this marketing cost from the revenues generated, we get an indication of the profitability of the product in the market area. Now I presume a display of these marketing costs and profit indicators classified by product and market area would constitute the basic data you'd use as input to your marketing decisions.

[Once more Sam has used the summary technique.]

R: That's right. If we could get just one display like that, it would undoubtedly reveal many instances in which we could significantly improve the effectiveness of our marketing efforts.

But that's just a beginning. Maintenance of such displays over a period of time might disclose hitherto unanticipated trends that can also be capitalized on.

I: And the discovery of such trends might indicate some marketing experiments you might like to try, either on a simulated basis or in the real world. Is that right?

R: Sam, I think you've got it.

I: Pro, you've given me a wealth of information to chew on, and I've got to spend more time digesting it.

[Sam now believes he has the information he set out to collect, and he is, therefore, moving to bring the interview to a close. However, first he checks to be sure Pro also believes they've reached their objective.]

I: So let me review what we've done, and let's see if we've really accomplished what we set out to do. My whole purpose in visiting you, Pro, was not to achieve anything in particular as far as our plans here at Hurd are concerned, but simply to get a firm grasp of the general concept of marketing productivity as it applies to Hurd or any other marketing organization. The object, of course, is that this conceptual understanding will guide us in data processing as we continue to cooperate with your department in our investigation of the feasibility of using automated techniques to improve our marketing productivity. Now, is there anything else about the general concept of marketing productivity that you think we should cover before we break off?

(A pause occurs while Pro thinks.)

R: Hum . . . no, I don't think so, Sam. This is probably a good place to stop. (Looks at watch.) I see I'm already ten minutes late for my next meeting. Frankly, I enjoyed speaking with you so much I lost track of the time. If there's any other information I can give you, please be sure to come back to me.

I: Thank you, Pro. To tell you the truth, I enjoyed it, too. And I certainly appreciate your offer, which I'm sure I'll take you up on—I'm bound to find things I still don't understand when I start reviewing my notes.

[Sam has moved to confirm that the door is open so he can come back and clarify anything he may have misunderstood or on which he has gathered insufficient data.]

R: Anytime, Sam.

[Now Sam moves to get Pro to commit to reviewing his interview notes.]

I: And just one other thing, Pro. When I get finished with my review of my notes, I'm going to write up what you told me. May I ask you to review the write-up?

R: I think I can swing that, Sam. You know, I've always wanted to get these things down on paper. Maybe before we're finished, you and I will have an article we can get published.

I: That's possible, Pro. Well, thanks for your time. I really appreciate it.

R: You're entirely welcome, Sam. See you shortly.

I: You bet.

[End of interview.]

The notes an interviewer can take during an interview are at best sketchy. In fact, he should take no more notes than necessary; these should consist of key words that will jog his memory as to what the respondent said. (Even symbolic representations can be used. For example, if the respondent makes a point on a subject he's addressed earlier in the interview, the interviewer can record just the new information and then draw an arrow back to his previous notes on that subject.) During the interview, the interviewer should spend as much time as possible listening to and looking at the respondent and as little time as possible writing.

However, as soon as possible after the interview, the interviewer should document the results of the interview. It's recommended that no more than 24 hours be allowed to elapse between the end of the interview and the documentation of the results. Otherwise, the interviewer will begin to forget what the respondent said.

A copy of the interview report should be sent to the respondent so he can review and approve it.

As an example of what an interview report should look like, here is the interview report for the interview just presented.

Interview Report

Marketing is those business activities that cause the flow of goods to customers.

However, in addition to generating sales, marketing should also produce information that leads to increases in marketing productivity.

Marketing productivity is the ratio of profits to marketing costs.

All other things being equal, the general relationship between the marketing costs for a given product in a given market area (such as a customer class or a territory) and the profit realized from the sale of the product is as shown [Figure 4.1].

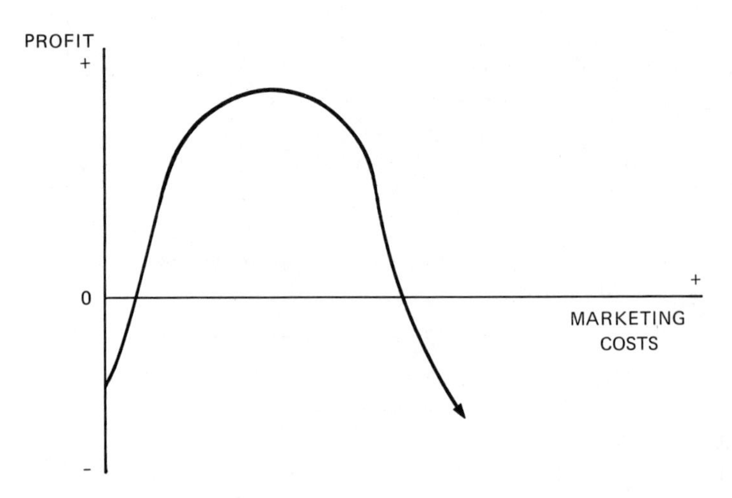

Figure 4.1

This relationship derives from the following situation.

1. When marketing costs are too low, not enough revenue is generated to cover all costs associated with the product.
2. As marketing costs increase, a progressively larger percentage of the total market for the product is captured, economies of scale are realized, and profits grow as the market is increasingly penetrated.
3. At some point, it becomes progressively more expensive to capture each added percentage of the market remaining, and as saturation of the market is approached asymptotically, marketing costs per added unit sold balloon to the point where all profits are consumed and larger and larger loss is generated.

Thus if the goal is to optimize profit, then for each product in each market area there is an optimum amount that should be spent for marketing. Marketing policy can err either by not spending enough (in which case the market is not adequately penetrated) or by spending too much (in which case too much effort is being made to expand the market penetration beyond its optimum point in terms of profit generated).

It thus becomes clear that, to be able to formulate an optimum marketing strategy for all products in all market areas, it is necessary to know:

1. The marketing cost for each product in each market area
2. The profit generated by each product in each market area
3. The relationship between items (1) and (2)

The relationships identified in item (3) are of interest both as they:

1. Exist now and have existed in the historical past, for all products in all market areas
2. Might exist if marketing costs were allocated among products and market areas in ways not previously tried

The second possibility implies that profit can be expressed as some function of marketing cost. In most cases, these functions can be determined only after experimenting with various marketing cost allocations.

However, determining these functions is a second order of business. The fact is that many organizations have no good idea of what their marketing cost and profit for each product in each market area are. If this information were available, the chances are enough instances of misallocation of marketing funds would be immediately recognizable to allow significant improvement of marketing operations without any attempt at experimentation to determine the precise relation between profit and marketing cost. In any case there would be little basis for directing such experimentation without first knowing current marketing cost and profit for each product by market area. Therefore the first order of business is to determine these costs and profits. A general approach to this determination is as follows.

Marketing costs are of two kinds.

1. Direct costs, which can be related to the marketing effort applied to each product in each marketing area
2. Indirect costs, which can't be so related

Indirect costs can be classified on a functional basis. For example, some functional areas in which indirect marketing costs are incurred are:

1. Inventory
2. Shipping
3. Order processing
4. Accounts receivable
5. Product planning
6. Packaging
7. Pricing
8. Advertising
9. Market research

These indirect costs must then be allocated to each product in each market area on the basis of a judgment as to what percentage of the indirect cost in each functional area can be attributed to the marketing effort for each product in each market area.

The sum of the direct marketing costs for a given product in a given market area and the indirect marketing costs allocated by function to the product in the market area constitutes the marketing cost for the product in the market area. Subtracting this marketing cost from the revenues generated by the product in the market area gives an indication of the profitability of the product in the market area.

A display of these marketing costs and profit indicators classified by product and market area may reveal many interesting situations. For example:

1. Certain customers or customer types may contribute to profitability; others may detract from it.
2. Certain product distribution channels may produce most of the profits; others may be minimally profitable or unprofitable.
3. Orders may have to reach a certain minimum quantity to become profitable.
4. Territories may vary widely in profitability.
5. Some products may be only marginally profitable or may detract from overall profitability.
6. The same product may be profitable in one package and unprofitable in another.

As indicated by these examples, product marketing costs and revenues must be broken down by a variety of market areas to provide a complete picture of marketing productivity.

A second step in optimizing marketing productivity without resort to market experimentation is to maintain displays of marketing costs and profits over a period of time. Trends that can be capitalized on may become evident in such displays.

A final word of warning. Marketing doesn't exist in a vacuum. Therefore a complete analysis of the impact of a change in marketing strategy should be made before such a change is instituted. For example, elimination of a marginally profitable product may make sense from a marketing productivity point of view, but it may also change the nature of manufacturing production runs in such a way as to increase unit costs for the remaining products to the detriment of overall company profitability.

BIBLIOGRAPHY

1. Kahn, Robert L., and Cannell, Charles F. *Dynamics of Interviewing,* p. 207. New York: John Wiley & Sons, 1963.
2. Kahn and Cannell, p. 241.
3. DeMasi, Ronald J. *An Introduction to Business Systems Analysis,* pp. 38, 39. Reading, Mass.: Addison-Wesley Publishing Company, 1969.

EXERCISES

1. What is interviewing?
2. What is the first problem the interviewer faces? How should he or she solve it?
3. Of what four things must the interviewer convince the respondent?
4. How does the interviewer start the questioning process? Why?
5. What is the best way for the interviewer to prepare for an interview?
6. Your assignment is to interview a Marcia Ericson, programmer, to determine the procedure she follows in doing her job. Develop an interview guide for this interview.
7. What are two ways in which the interviewer can get the respondent to expand on his or her remarks without asking a specific question?
8. Name three types of nondirective probes.
9. Describe three scales for rating the effectiveness of a probe.
10. The following are four interview situations. In each one, indicate what a good interviewer response would be and explain why.
    a. I: Do you expect sales to increase next year?
       R: Well, I sure hope so. If we have another year like the last one, we'll be in real trouble.
    b. I: What do you think of the order entry system?
       R: Oh, it's OK, I guess.
    c. I: Do you think we'll experience a significant increase in productivity in the next year or two?
       R: I doubt if we're going to hire many more people.
    d. I: Do you think the work in your department should be organized on a project basis?
       R: Well, over a period of time we do work on a variety of different jobs, and even over the life of any one of the jobs, the people needed to do the job fluctuates. But I'm not sure my people would live with such an unstable work environment.
11. If the interviewer must ask threatening questions, where in the interview should they be asked? Why?
12. What three things must the interviewer do when terminating an interview?

Part 2

COMMUNICATION

In addition to collecting data from the user the analyst must communicate to the user the information he has collected, analyzed, and synthesized, so the user can be sure that communication has occurred and that user and analyst are in step on system definition. As the analyst collects data, he must document these facts in a correct and communicative way. He must provide verbal and written feedback to determine that the information collected is, in fact, correct. In particular, he must be able to document previously undocumented procedures in a concise and meaningful way. He must prepare the functional specifications document and communicate its contents in presentations. He generally prepares the user manuals that are used to train user personnel and subsequently become the user's reference on use of the system. If it's the analyst who first sees that there's a better way to do things, he must communicate his ideas to the user: perhaps in documentation, perhaps in formal and informal presentations, perhaps in both. Thus Part 2 of this book is on communications skills.

We have two forms of formal communications: reports and presentations. For a discussion of writing reports and all forms of documents, we refer you to Gildersleeve, *Organizing and Documenting Data Processing Information* (Rochelle Park, N.J.: Hayden, 1977). Consequently Part 2 contains just one chapter.

5. Presentation Techniques

A special type of presentation in which the analyst is involved is the training of user personnel for their role in the production running of the data processing system to be installed. For information on how to prepare such a training program, see Robert F. Mayer and Kenneth M. Beach, Jr., *Developing Vocational Instruction* (Palo Alto: Fearon Publishers, 1967).

5

Presentation Techniques

The purpose of business communications is to convey information and to arrive at decisions.

The two types of business communications, written and verbal, are generally referred to as *reports* and *meetings*. We'll go along with this terminology as long as it's clear that:

> Whenever anyone commits something to paper, be it a memo, a letter, a note, what have you, that something is a report.
>
> Whenever two or more people get together to discuss something, that's a meeting.

Meetings are poor ways to convey information. This is demonstrated by the classic experiment of whispering a sentence into the ear of a person, who whispers it to another person, and so on, until finally someone whispers it to you in a form you don't recognize as the sentence you started with. The best vehicle for communicating information is the report, because each person must absorb information at his or her own pace. The report allows each reader to go fast or slow, to reread, to skip around, to do anything he or she finds useful.

On the other hand, the report is a poor vehicle for trying to arrive at a decision. If a decision must be made, an interchange of ideas and attitudes is needed, and if this is done by reports, the process will be so protracted as to make decision making unnecessary—whatever would have required a decision would have resolved itself, perhaps disastrously, before the decision could be reached. A meeting, on the other hand, brings together the people concerned with the impact of the decision and allows them to reach a mutually acceptable solution in an environment conducive to a rapid interchange of ideas and interests.

Of course, decisions shouldn't be reached without considering the pertinent information, and thus meetings always involve some interchange of information. Nevertheless, the best approach to organizing a meeting is to document all the facts that seem to bear on the decision to be made and to distribute these reports to the attendees long enough before the meeting to let them absorb the facts in advance. As a result, during the meeting the attendees can concentrate on the business at hand—the decision to be made.

If you call a meeting, it should be because you need a decision on some matter by the people you invite. The matter on which you need the decision is your *problem,* and generally you have a solution for the problem and want the meeting members to approve your decision. These facts determine the form of the meeting (which you can also think of as your presentation).

5.1. PRESENTATION FORMAT

A meeting begins with a problem statement. Agreeing on an approach to solving the problem is the purpose of the meeting. The problem statement contrasts what exists with what the meeting members want to exist.

The best way to reach agreement is to start at a common point and collectively work toward a common destination. Thus, after agreement is reached on the nature of the problem to be solved, the next step is to find an area of agreement among all the meeting members. This area of agreement will be in terms of the beliefs and attitudes the members share.

Once the area of common agreement has been established, the rest of the meeting agenda is fixed. First comes the presentation of the logical sequence of steps that connects the area of common agreement with the proposed solution to the problem. Then a request is made for endorsement by the meeting members of the proposed solution. When the endorsement is received, the meeting is closed with an outline of the steps to be taken as a result of the agreement.

5.1.1. Problem Statement

A problem can be defined as a disparity between what exists and what we want to exist. To get your audience's attention, you must describe your problem in terms of what now exists as opposed to what *they* want to exist. You must also make it clear that there's a solution to the problem—a way for the audience to get from the present state to the desired state. Both these aspects of problem statement are vital. If the problem isn't something they want to solve, you'll never get their attention in the first place. But having gotten their attention, you must then assure them that there's a solution. If you don't, you're just drawing their attention to what is to them a frustration, and instead of getting their cooperation, you'll irk them.

5.1.2. Proposed Solution

Not only must you assure your audience that there's a solution; you must have a solution with which to present them. Your solution must be clearly worked out in complete detail. Otherwise, there's no point to your presentation, and you'd be better off not calling the meeting. Without a solution to present, you're just postponing the moment of frustration until the end of the meeting, and the longer the meeting, the greater the antagonism you will arouse.

5.1.3. Developing the Presentation

You've opened your presentation with the problem statement. You know what your proposed solution is. The purpose of your presentation is to assure that when you propose your solution, you'll get agreement. In other words, when it comes time to make your proposal, you must have your audience with you.

The best way to do this is to start with your audience at a common point and then carry them with you to your conclusion. This means that the presentation between the problem statement and the proposed solution consists of two parts.

1. Establishing a common ground between you and your audience
2. Demonstrating that your proposed solution follows inevitably from this common ground

You must start where your audience is. That means you must know your audience—their desires, fears, areas of competence, limits of competence, and attitudes toward you and your goal. On the basis of this knowledge you must determine the area of common agreement between you and your audience.

It's generally not possible to move directly from the area of common agreement to your proposed solution. Instead, you must carry your audience point by point through a series of logical steps that connect the area of common agreement with your proposed solution.

For example, suppose you want to convince your company that the first phase in the development of a data processing system is to have the data processing department and the user come to an agreement as to what the system is going to do; that this agreement should spell out these objectives in detail; and that the agreement should be in writing. This is the action you want your audience to take, and the logical steps leading to this conclusion might be as follows.

1. Begin by establishing a problem. This might be that the company's data processing system development projects have consistently been subject to large overruns of both time and money. If this is truly a problem, it should be easy to establish the problem's existence with historical data.
2. Now you must find the common ground on which your audience stands. This might be the belief that such overruns are undesirable.
3. Finally, from this common ground, you must convince your audience of the inevitability of your proposed solution. The development might be as follows.
   a. To make reliable time and cost projections, a detailed plan of what resources are going to be used, and when, must be developed.
   b. To develop such a detailed plan, a firm base consisting of an agreement as to what is to be done must exist.
   c. Therefore . . . (your proposed solution now follows).

Notice that the steps in the development have an inherent logical order. If you want your audience to accept your proposed solution, you must present your points in a sequence that makes sense to them. This suggests a presentation technique that is known as the *wrap-up technique* and has the following structure.

> Make sure of your common ground by stating it explicitly and getting your audience's acceptance. Then go on to the first point and develop it just far enough to get your audience's acceptance. Then go to the next point, and so on, until your audience has accepted your proposal. At each point along the way, spend just as much or as little time as is necessary to gain your audience's acceptance.

Notice that the wrap-up technique has it's own built-in pacer for the presentation. You go just as fast as your audience will let you, wrapping up the points one by one and spending only as much time on each point as is necessary to gain acceptance.

Unfortunately, every rose has a thorn or two. The disadvantage of the wrap-up technique is that presentations are generally made to groups, and some of the individuals in the group will grasp your points more readily than others. Thus the major challenge you will face when using the wrap-up technique is of convincing the members of the audience who are slower to agree without losing the interest of those who already agree to a particular point.

The approach here is to enlist the aid of those who agree with you in convincing those who don't. Get them involved. Not only may they be able to make your point in a more telling way than you've been able to; they will also be selling themselves more firmly on your point.

In case you find it impossible to convince one of the members of your audience, you must then choose between losing the individual and losing the rest of the audience. If you don't want to lose the individual, then you must stay on the point to which he objects, because if he doesn't agree with this point, he won't agree with your solution either. If you're willing to lose the individual to maintain the attention of the rest of the audience, then you go on to your next point.

How do you make this decision? You go back to the goal of the presentation, which is to get your proposal adopted. If you can't reach this goal without the dissenting individual's acceptance, then you must fight it out on that point no matter how long it takes, because this particular battle is crucial to the outcome of the war. If you can afford the loss of the individual's support and you see that the rest of your audience is becoming restive, you have little choice but to go on.

5.2. METHOD OF PRESENTATION

Once you've settled on the form of your presentation, the next question is, How are you going to make your presentation? The fundamental problem of presentation is getting and maintaining your audience's attention.

If you've stated the problem in terms of what your audience wants and have indicated that the problem can be solved, you'll have no difficulty getting your audience's attention—it will be riveted on what you're going to say next. Your challenge is to maintain this attention through the remainder of your presentation.

The fundamental rule for maintaining the audience's attention is: *Be brief.* This is the case for two reasons.

1. The shorter the presentation, the shorter the period in which you have the problem of maintaining attention.
2. Brevity appeals to your audience. As H.M. Boettinger says in *Moving Mountains* (New York: Macmillan, 1969), "Nothing gives an audience greater pleasure than the experience of quickly grasping a new idea."

The best way to get a point over to your audience rapidly is by using examples. Examples bring abstract ideas down to particular cases with which your audience can identify.

Other devices for maintaining attention are:

1. Team presentations
2. Visual aids
3. Avoiding the use of jargon

5.2.1. Team Presentations

If you have a complex subject that's going to take a long while to develop, the use of a team of presenters solves the problem of the monotony introduced by exposing your audience to one person for a long period of time. Any presentation requiring more than an hour should probably adopt a team format.

While a team presentation provides variety, it creates special problems for you. You must be sure that each team member knows his job and is capable of carrying it out. You must also be prepared to coordinate the presentation. You must open and close the meeting, and you probably should say a few words between each presentation to define and clarify the relation of the presentations to each other and to the meeting as a whole. You must also monitor the meeting to be sure it stays on track and be prepared to step in if it shows signs of bogging down or veering from the ultimate goal.

5.2.2. Visual Aids

Too much has been said about visual aids already. When used effectively, visual aids enhance a presentation significantly. However, visual aids are just a tool. Using them in a presentation doesn't guarantee success. That people have lost sight of this fact is amply shown by the ubiquity of the boring presentation that consists of an interminable series of overhead transparencies shown one after the other.

A successful presentation is one in which the audience recognizes a problem they want solved and agrees with your proposed solution. Visual aids contribute to this success only if they're appropriate.

There's a school of thought that says that every visual aid should be self-explanatory and stand by itself; that is, your audience should understand it and grasp its significance unaccompanied by any verbal presentation. This position is sometimes extended to the idea that each member of the audience should be supplied with a copy of each visual aid and that this collection of visual aid copies should contain all the information in the presentation. This is the kind of thinking that leads to end-to-end, detailed visual aids. We thoroughly disagree.

In many instances, handing out copies of material is an improvement on the use of a visual aid. It reduces note taking and gives everyone their own reference. But a visual aid is what its name says it is: an aid. It should support and contribute to your presentation, not substitute for it. When your presentation refers to charts or tables, visual aids are the only effective way to get your point across. A checklist of main points to keep your audience oriented may also be an appropriate visual aid. But each visual aid you use should help you make your presentation. If it only duplicates your presentation, you're better off without it. You want your audience to concentrate on you, not your flipchart.

Here are a few rules of thumb for constructing visual aids.

1. Visual aids done by artists are nice, but far from indispensable. You can construct your own visual aids quite competently.
2. You have basically three choices for visual aids.
   a. *The real thing.* For example, if you're advocating the use of a piece of equipment, having the actual equipment to point to is a decided advantage.
   b. *Exhibits.* Large charts are an example. The most common example is the flipchart, and unless a separate chart is already available, this is probably to be preferred. It keeps your visual aids organized.
   c. *Projections.* Common examples are overhead projections, slides, videotapes, and film. These are the next best thing to the real thing. They're also good for complicated charts that are difficult to draw on a large scale. However, they require fiddling with equipment, which is a bother to you and your audience. Consequently, given a real choice, you should probably opt for a flipchart. However, if you have to take your show on the road, a set of transparencies is easier to carry and less likely to be lost than a flipchart.

3. If you're making your own visual aids, hand-letter them. Typing doesn't project well. Use capital block letters.
4. Make sure your visual aids are visual—be sure they can be read. This applies to everything, not just the major headings. The smallest detail should be clearly visible from the back of the room. Don't trust to luck; check the visibility of your aids by putting them where you're going to use them and then going to the back of the room to see whether you can read them easily.
5. Put an identifying heading on each visual aid. Unlabeled visuals are confusing.

5.2.3. Avoiding the Use of Jargon

Generally, the reason you're presenting a problem and a proposed solution is that they fall within your area of expertise. You probably know more about the subject than your audience, if only because you've been concentrating on it more than they have. On the other hand, your audience typically is concerned with something other than your area of specialization. You need their concurrence because your proposed solution impinges on their areas of activity.

You and your audience have an overriding commonality of goals. That's why you're talking with them. But you collectively pursue your common goals by specializing on various aspects of their attainment.

Typically, you're separated from your audience by your specialty. And such a situation lends itself to the use of jargon—the language of your specialty—when communicating with people who don't share that specialty.

Jargon should be avoided for several reasons.

1. It doesn't aid communication. When you use jargon, people don't know what you're talking about.
2. Use of jargon is a status symbol. It emphasizes that you're something your audience isn't. The purpose of a meeting is to make a decision, decisions are made among equals, and to inject status into a decision-making situation is to create antagonism and resentment. You may have a recommendation decidedly in your audience's interests, but if you alienate them, they'll withhold their endorsement out of spite.
3. Use of jargon gives your audience the impression that you're not telling them what they need to know. This is antithetical to the goal of your presentation, which is to keep your audience with you.

Nevertheless, to get your point across, it may be necessary to draw on your special knowledge and share it with your audience. How do you do this without using jargon?

The first step is to be sure you understand what your audience needs to know and to then confine yourself to the information thus identified. This information is often minimal and frequently involves a distortion of the complex of facts as you in your specialty know them. It may be disheartening to realize how little of your knowledge your audience really wants in order to make their decision. Nevertheless, if it's their approval you want, restrict yourself to what they have to know. Remember: Be brief.

If some special knowledge is essential to the presentation, your best bet is to present it in the form of an analogy to something with which your audience is familiar. Again, this may be discouraging, because analogies by their nature always distort. But if the analogy, distorted though it may be, puts across without distortion the kernel of information your audience requires, then it represents your best chance to achieve communication. If you cannot find an analogy and must use some terms of your trade, then define them in words your audience does understand.

Numbers are a special type of "jargon" that must often be dealt with. Both small numbers and large numbers are difficult to put across.

1. Decimals represent a degree of precision that repels many people. Therefore:
   a. If the numbers with which you're dealing consist of integer and decimal parts, round the numbers and drop the decimals if at all possible.
   b. If the numbers are pure decimals, use fractions instead.
2. People have trouble relating to large numbers. Try to express them in terms with which people commonly deal. The classic example is expressing the national debt in terms of the amount owed by each citizen.

5.3. AVOIDING DISTRACTIONS

You want your audience's single-minded attention, and this is difficult enough without providing temptations for the mind to wander. These are the distractions to avoid.

1. *Noise.* The ideal is for the only sound to be that related to your presentation. This is a difficult ideal to achieve, but you

should come as close as possible. The first big step is to choose the quietest possible location and time for your presentation. The second step is to remove the noises that are present. For example, if the chairs squeak, get them oiled. If there is a telephone in the room, take the receiver off the hook.

2. *Sight.* Aside from the essentials (chairs, tables, lighting fixtures, etc.), the only things in the room should be related to your presentation. Get the visual aids from yesterday's presentation out of the room. If the room has windows, draw the drapes or pull the blinds.
3. *Discomfort.* An uncomfortable person is a distracted person. Chairs should be comfortable. The temperature shouldn't be so warm as to encourage drowsiness or so cold as to make people want to leave the room. Lighting should be sufficient but not glaring. People shouldn't have to crane around pillars to see visual aids.

All this detail isn't very inspiring, but if the success of your presentation is important to you, you should allow yourself the time before your presentation to check into these matters and arrange to avoid them.

One last point: The object of holding a meeting is to allow people to exchange ideas, attitudes, and information, so they can agree on a decision. Room layout should encourage such an interchange among equals. Consequently the best room layout is conference room or "U" style.

5.4. YOUR ATTITUDE

A large part of the success or failure of your presentation depends on the attitude you have toward your audience. Your attitude is something you can't hide from your audience; it makes itself evident through your unconscious actions and choice of words. The only approach you can take that will help you in this important area is to develop the right attitude.

The first point to recognize is that the purpose of your meeting is to come to a decision, and decisions are made among equals. You must treat the members of your audience as your equals. This doesn't mean you should subject them to uncalled-for familiarity; they should be treated with dignity and tact. Nor does it mean that the members of your audience aren't different from you; they are, both in background and in interest. It *does* mean that your attitude should be neither one of superiority nor one of submissiveness. An air of superiority creates antagonism and resentment, which are not conducive to keeping your audience with you. Submissiveness is fatal, as we show in detail next.

When making a presentation you're selling an idea—advocating a change. As Boettinger points out, ideas are always a buyer's market. Your idea is just one of many on which the members of your audience are continuously requested to risk their resources and reputations. And as Machiavelli has pointed out,

> There is nothing more difficult to carry out, nor more doubtful of success, nor more dangerous to handle, than to initiate a new order of things. For the reformer has enemies in all those who profit by the old order, and only lukewarm defenders in all those who would profit by the new order, this lukewarmness arising partly from fear of their adversaries—and partly from the incredulity of mankind, who do not truly believe in anything new until they have had actual experience of it.
>
> from *The Prince*

Given this situation, no audience is going to risk its fortunes by endorsing an idea proposed by a person motivated by fear, which is the light in which a subservient attitude puts you. You mustn't appear arrogant, but you must appear confident.

The best way to appear confident is to be confident. Confidence in an idea springs from two sources.

1. *Deep personal commitment.* You must believe totally in the proposal you're going to put forward for adoption.
2. *Genuine knowledge.* You must understand the situation in which you're making your proposal and be competent in all areas bearing on it.

If you don't meet both these qualifications, you're better off not making your presentation. Even if you're committed and competent, look before you leap. Try your ideas out on the inquiring and skeptical minds of your colleagues before attempting to sell your ideas to the powers that be.

Many of the questions and comments from your audience will strike you as unintelligent or misinformed. This is never the case. These questions and comments arise from two sources.

1. Your presentation consists primarily of words, and words are ambiguous. Your words can be interpreted to convey your

meaning, but frequently other interpretations are equally possible.

2. Your audience isn't a mechanical receptor, but a varied collection of individuals, each with his own hopes, aspirations, fears, and concerns. What may seem to you a completely objective statement of fact may appear, to one or more members of your audience, as a decided threat.

Therefore your response to all remarks by your audience should be courteous, serious, and good-natured, as befits one equal dealing with another. If a person misunderstands what you say, view it as an opportunity to help him. Relate what you're trying to say to his own background and experience. Demonstrate your ability to be flexible in dealing with people of different backgrounds. Treat a demonstration of fear as an opportunity to give reassurance. At the least, demonstrate that you have understanding of and sympathy for the person's concerns.

5.5. HANDLING QUESTIONS

The object of your presentation is to carry your audience from the area of common agreement to your proposed solution. You have no way of knowing whether you're accomplishing this unless your audience participates in the presentation. Successful presentations aren't monologues, but dialogues between you and your audience.

The most common way for a member of your audience to participate in your presentation is to ask a question. An attitude commonly held by presenters is that questions are hindrances, roadblocks thrown in the way of your smooth presentation, and frequently the questioner is attributed with hostile motives. While this may occasionally be the case, in general nothing could be further from the truth.

One thing is certain: Asking a question takes effort. It's a certain sign that the asker is paying attention, which is what you want. Therefore questions from the audience indicate the health of your presentation, not its sickness.

In most cases the questioner is with you, not against you. His motive for asking the question may vary. Perhaps he agrees with you but senses that some other members of the audience are silent but unconvinced and wants you to bear down harder on your point. Perhaps his agreement is tentative and he wants reassurance. If he adopts your proposal, he may have some selling to do at home, so perhaps he wants more detail to help him out. Perhaps he simply doesn't understand and wants some clarification.

Generally, if one person asks a question, other members of the audience have the same question. Thus a question usually represents an opportunity to inform and gain the commitment of many members of your audience, not just one.

Even if a question is motivated by hostility to your position, you should be glad it is asked. Sooner or later you must deal with this hostility, and the question affords you the opportunity to contend with it in the presentation. If the hostility is ill-founded, the question gives your supporters in the audience the chance to show that it is unjustified. If your position is well-taken, you won't have to defend yourself against hostility—your audience will do it for you. If the hostility is well-founded, perhaps your logic is faulty, and the time to find out about this is during the presentation, not after your proposal is adopted and runs into trouble.

In preparing your presentation, try to anticipate the questions and objections your audience may raise and develop good responses to them. Such an exercise tests the viability of your ideas and protects you against embarrassment and the consequent loss of momentum in your presentation.

There may be questions to which you don't have an answer. If this happens, acknowledge the situation openly. Any quibbling or attempts at evasion reveal weakness, which lowers your audience's willingness to entrust themselves to your proposal.

You should turn a question you can't answer over to the audience. If someone in the audience can answer it, everyone benefits. If no one can, this shows that your lack of an answer stems from the general state of affairs rather than from a failing on your part.

5.6. PREPARATION

Things never go exactly as anticipated. To minimize the probability of unanticipated developments, try your presentation out on a practice audience before actually giving it. The chances are such an audience will be friendly, so get at least one person in the audience who is willing and able to play the role of devil's advocate.

5.7. THE INVITATION

To maximize the probability that you'll have in your audience the people you want, you should invite them to your presentation early enough to give them the chance to:

1. Arrange their schedule to attend
2. Prepare to participate actively and constructively

If a person is going to take the time out of his schedule to attend your presentation, you owe him the courtesy of telling him the presentation's purpose. (From a more selfish point of view, you should realize that the more influence a person has on getting your proposal adopted, the less likely he is to attend a presentation whose purpose is unknown to him.) For example, if the purpose of your presentation is to propose the adoption of a procedure for nailing down what a data processing system is going to do, and to argue that such adoption will significantly curtail project overruns, then the invitations you send out to your intended audience should say so.

Always remember that the purpose of your presentation is to obtain your audience's acceptance of your proposal. Therefore you want your presentation to be concerned with the pros and cons of your proposal, not with the details that make up your proposal. The most desirable thing is to have your audience educated on these details before the presentation begins, and you should give them every opportunity to achieve this level of sophistication by supplying them with these details when you extend the invitation. Thus, to continue the above example, your invitation to your presentation should be accompanied by a document that details:

1. How the user and the data processing department are to come to an agreement as to what the data processing system is going to do
2. What form this agreement will take
3. The level of detail incorporated into the agreement

Of course, the invitation should also state that the presentation will be made on the assumption that the material accompanying the invitation has been absorbed.

5.8. SUMMARY

Here is a checklist for preparing a presentation at a meeting.

1. Have you stated the problem—what exists versus what the audience wants to exist?
2. Do you have your proposed solution clearly in mind—do you know what you want the audience to do at the end of your presentation?
3. Are you personally committed to your proposal—do you believe in what you want your audience to do?
4. Are you competent to make the planned proposal—do you know what you're talking about?

5. Do you know your audience—their desires, fears, areas of competence, limits of competence, importance to your goal, attitudes toward you and your goal?
6. Have you identified the area of common agreement between you and your audience?
7. Have you developed the sequence of points that connects the area of common agreement with what you want the audience to do?
   a. Have you kept the points to a minimum?
   b. Have you restricted the technical material to what the audience must know? Have you expressed these points in a way your audience can understand, either by analogy or by definition?
8. Have you tried out your ideas on the inquiring and skeptical minds of your colleagues?
9. If this will be a long presentation, do you have a team to make it?
   a. Does each team member know his job, and is he capable of carrying it out?
   b. Are you prepared to coordinate the presentation?
10. Do you have a trenchant example with which to drive home each point?
11. Are you prepared to involve your audience in the development of your points? (Remember: If the audience doesn't participate, you have no way of knowing whether they agree with your development.)
12. Have you used visual aids whenever they can help in making your points?
    a. Are they large enough to be read easily?
    b. Do they have headings?
13. Is the room free of distraction? Will its layout encourage discussion?
14. Are you prepared to treat the members of your audience as your equals, neither better nor worse, just different?
    a. Are you willing to listen courteously to all comments?
    b. Are you willing to do your best to answer every question seriously and with good nature?
15. Have you anticipated the questions and objections your audience may raise?
    a. Do you have a good answer for each one?
    b. If you don't have an answer to a question, are you willing to admit it? (And don't forget to ask your audience tactfully if they have an answer to the question.)
16. Have you rehearsed your presentation? Was there a devil's advocate in the audience?

EXERCISES

1. What are the purposes of business communications? What forms of communication are used to effect these purposes?
2. You have some information you want to convey. What should you do: write a report or call a meeting? Why?
3. You want Charlie to make a decision. Should you: write him a report or meet with him? Why?
4. Are there situations where you'd both write a report and call a meeting? If so, when?
5. What is the general meeting format?
6. What should you do if you don't know the answer to a question asked during a presentation?
7. How should every good presentation end?
8. What are the three essential parts of a good presentation?
9. What is the essense of a problem statement?
10. Why should you encourage your audience to make observations and ask questions?
11. What should every good visual aid have?
12. How big should a visual aid be?
13. What is the best way to make a point?
14. How can you avoid jargon?
15. Where must the development of every good presentation begin?
16. How would you organize a long presentation to make it more effective?

Part 3

THE FUNCTIONAL SPECIFICATIONS

The major document prepared by the analyst is the functional specifications. Consequently Part 3 of this book is on the functional specifications.

Part 3 begins with Chapter 6, "The Functional Specifications," which spells out detailed standards for the form and content of functional specifications.

As we have already mentioned, from the user's viewpoint a major aspect of the proposed system is how he and his staff will interface with the system. Consequently the functional specifications describe in detail the system's input and output forms and the dialogues at the terminals, if any. The designer is concerned with forms and terminal dialogue and does receive training on these subjects. However, the emphasis in this training is on how the forms and dialogue relate to the structure of the system. The analyst is concerned with how people relate to the form and dialogue—the human engineering of the system. Chapter 7 and 8, in Part 3, are concerned with this human engineering of forms and dialogue.

As we've also said, the user is concerned that the proposed system carry out the procedures for which the user is responsible. Consequently the functional specifications also describe these procedures in detail. In the belief that the best way to document these procedures is with decision tables, we devote Chapter 9, the last chapter in Part 3, to decision tables.

6

The Functional Specifications

The functional specifications document is the physical manifestation of the agreement reached between the data processing department and the user as to what the system is to do. The user and the data processing department come to an agreement on three aspects of the data processing system.

1. System description
2. Acceptance criteria—criteria for acceptance of the system on completion of development
3. Measuring benefits realization—the method to be used during system life to determine the extent to which the system realizes the benefits specified

Since the functional specifications document spells out agreements between the user and the data processing department, it's written in such a manner that:

1. Whenever a question of the form, "What did we say was going to happen in this situation?" arises between the data processing department and the user, the natural reaction is to look for the answer in the functional specifications.

2. Whenever it becomes desirable to change system function in a small or large way, whether the suggestion for the change is initiated by the user or the data processing department, the natural way to record the change is to modify the functional specifications.

These criteria represent an ideal. No functional specifications you write will meet the ideal perfectly. Questions will arise on which the functional specifications won't shed light. However, the fact that the ideal isn't attained doesn't negate its desirability.

You write the functional specifications to be understood by the user. This means you write them:

1. In the user's words
2. In terms of what the system is going to do for the user

These points are worth some expansion.

6.1. IN THE USER'S WORDS

Any application area has its own nomenclature, or terminology. It's in the language of this terminology that you write the functional specifications. If you're concerned that as a result the specifications document will lose some of its comprehensibility within the data processing department, you can append to it a glossary in which unfamiliar terms are defined. If the functional specifications are written in this language, it is more likely that:

1. You understand what the user is talking about
2. The user understands what he's getting

6.2. WHAT THE SYSTEM IS GOING TO DO

The functional specifications are a description of what the system is to do and how the user will go about getting it to perform its functions. The functional specifications describes:

1. Every system operation that can be performed in response to input introduced by the user. This includes the description of any output that may be produced
2. Each possible input the user can introduce
3. Controls

A description of the contents of functional specifications raises the question of how much detail is necessary. The only possible answer is a functional one. As we have mentioned, the functional specifications document is the physical manifestation of an agreement between the data processing department and the user. To be a party to this agreement, the user must understand what he's agreeing to, specifically:

1. What he's getting
2. What he's not getting
3. What his obligations are

Therefore you write the functional specifications in sufficient detail so the user can understand them.

6.3. THE CONTENT OF FUNCTIONAL SPECIFICATIONS

As we've already said, the functional specifications document spells out agreements with respect to:

1. System description
2. Acceptance criteria
3. Measuring benefits realization

The rest of this chapter is devoted to a description of these agreements.

6.3.1. System Description

The description of the system is subdivided into five parts.

1. Processing
2. Output
3. Records
4. Input
5. Controls

The sections appear in the description in the order listed. The output, records, and input sections could almost be considered appendices of the processing section and are used by the processing section in this way. Consequently, in this chapter we'll first describe the output, records, and input sections of the system description. Then we'll take up the description of the processing section. Finally, we'll cover the description of the controls section.

6.3.1.1. Output. This section describes the content of each output, how it's to be presented to the user, estimated output volumes, and time schedules. For example, if the output is a printed report, the final form doesn't have to be defined, but the meaning of each field on the report and the general form layout are spelled out. Or if the output is a display on a remote device, the nature of all displays doesn't have to be described, but a listing of possible displays, their meaning, and the general characteristics of their format do appear.

For example, if the output is to be a check and a check stub, a description of this output might be as follows.

1. The check will fit into a standard business envelope.
2. The check stub will be the same size as the check.
3. The check will contain the following information.
   a. Employee number (nine characters)
   b. Date, in numeric form: month (two digits), day (two digits), year (two digits)
   c. Name (twenty-one characters)
   d. Net pay (seven digits, to the nearest cent)
4. The check stub will contain the following information.
   a. Date (same format as on check)
   b. Department number (four characters)
   c. Employee number (nine characters)
   d. Gross pay (seven digits, to the nearest cent)
   e. Net pay (seven digits, to the nearest cent)
   f. Federal withholding (six digits, to the nearest cent)
   g. State withholding (five digits, to the nearest cent)
   h. FICA tax (five digits, to the nearest cent)
   i. Retirement plan contribution (five digits, to the nearest cent)
   j. Insurance contribution (five digits, to the nearest cent)
   k. Twelve other deduction fields, each of the following format
      i. Deduction identification code (two characters; all code definitions are preprinted on the stub)
      ii. Amount (five digits, to the nearest cent)

6.3.1.2. Records. This section defines every field in the files. For example, in a payroll system a record for an employee might be described as follows.

1. Identification information
   a. Badge number (eight characters)
   b. Name (twenty characters)

c. Address (thirty characters)
d. Social Security number (nine characters)
2. Descriptive information
a. Hourly rate (five digits, to the nearest tenth of a cent)
b. Number of income tax exemptions (three digits)
c. Union dues key (one character)
d. Blue Cross key (one character)
3. Accumulations
a. Gross earnings, year to date (seven digits)
b. Withholding tax, year to date (seven digits)
c. FICA
i. Year to date
(1) Taxable earnings (seven digits)
(2) Tax (five digits)
ii. Quarter to date
(1) Taxable earnings (seven digits)
(2) Tax (five digits)
d. Days of allowable absence remaining, annual (three digits to the nearest tenth of a day)—this figure will be cumulatively reduced by days of absence and will become negative if absence exceeds allowance
4. Bond information
a. Size of bond key (one character)
b. Weekly bond deduction amount (five digits)
c. Cumulative bond deductions (five digits)

If your installation has a data dictionary, using the dictionary to generate this section of the functional specifications has several advantages.

1. The section format is standardized.
2. Sufficient care is taken to get field definitions meaningful to the user into the data dictionary.
3. The discipline imposed by the dictionary is applied to the project at an early stage.

6.3.1.3. Input. This section describes the content of each input, how the user is to enter the input into the system, estimated volumes of input, and time restraints under which the input must be entered. For example, if input is a form the user fills out, the final form doesn't have to be defined, but the meaning of each possible entry on the form and the way in which it's to be entered are spelled out. Or if the input is to be keyed in on a remote device, the exact nature of all key-ins doesn't have to be described, but a list of possible key-in types, their meaning, and the general characteristics of their format do appear. For example, if the input is to be an application for a small group insurance policy, it might be described as follows.

1. Agent's name (twenty-one characters)
2. Agent's street address (twenty-one characters)
3. Agent's city address (thirteen characters)
4. Agent's state address (two characters)
5. Agent's zip code (five digits)
6. Agent's code number (five characters)
7. Second agent's name (twenty-one characters)
8. Second agent's code number (five characters)
9. Agency manager name (twenty-one characters)
10. Agency manager code number (three characters)
11. Agency code (two characters)
12. Application number (six characters)
13. Policy holder name (twenty-one characters)
14. Policy holder street address (twenty-one characters)
15. Policy holder city address (thirteen characters)
16. Policy holder state address (two characters)
17. Policy holder zip code (five digits)
18. Rate classification (three characters)
19. Life insurance (choose one)
    a. Flat plan
        i. Amount (five digits, to the nearest dollar)
    b. Percentage of earnings
        i. Percentage (choose one)
            (1) 100%
            (2) 150%
            (3) 200%
20. Weekly indemnity (choose one)
    a. No
    b. Yes
        i. Amount (three digits, to the nearest dollar)
        ii. Maximum duration (choose one)
            (1) For states with no statutory plans
                (a) 13 weeks
                (b) 26 weeks
                (c) 52 weeks
                (d) 104 weeks
            (2) For states with statutory plans
                (a) 26 weeks
                (b) 78 weeks

21. Accidental death and dismemberment (choose one)
    a. None
    b. Same as life amount
    c. Twice life amount
22. Dependent life (choose one)
    a. No
    b. Yes
        i. Spouse amount (choose one)
            (1) $1,000
            (2) $2,000
            (3) $3,000
            (4) $4,000
            (5) $5,000
23. Health insurance (choose one)
    a. No
    b. Yes
        i. Type (choose one)
            (1) Plan 1
            (2) Plan 2
            (3) Plan 3
            (4) Plan 4
            (5) Plan 5
        ii. Coverage (choose one)
            (1) Participant only
            (2) Participant and dependents
        iii. Dental (choose one)
            (1) No
            (2) Yes
        iv. Maternity (choose one)
            (1) None
            (2) $400
            (3) $600
            (4) $800
        v. Medicare patients (choose one)
            (1) No
            (2) Yes
24. For each participant
    a. Name (twenty-one characters)
    b. Date of birth, in numeric form: month (two digits), day (two digits), year (two digits)
    c. Sex (choose one)
        i. Male
        ii. Female
    d. Annual earnings (six digits, to the nearest dollar)

e. Dependent status (choose one)
   i. None
   ii. Spouse only
   iii. Children only
   iv. Spouse and children

In general, the more clearly defined the input and output are, the better are the chances of delivering the system to a satisfied user.

6.3.1.4. Processing. This section describes, without the use of data processing terminology, the processing the system is to do. It describes what processing the system is to do for the user, not how the system is to do the processing.

The description is carried out to the extent necessary to define the way each output and record field is derived. In many instances the name of the field is sufficient to describe its source—for example, vendor name or employee badge number. In other instances more detail is necessary—for example, the algorithm used to determine reorder quantity or the priority with which deductions are to be applied to an employee's wages.

The point at which some functional specification writers balk is the extent to which a description of the processing operations done by the system is required. However, most of this information already exists. It doesn't have to be rewritten. All that's necessary is a reference to the appropriate document. For example, rules for computing wages are specified by citing the appropriate labor agreement, methods for computing withholding tax are described by referring to the appropriate government document, and the algorithm for computing reorder quantity in a new order system can generally be found in correspondence or reports documenting the development of the algorithm. However, for the sake of completeness all such cited documents should be attached to the functional specifications as appendices, and should be referred to as such in the appropriate places in the processing description section.

We suggest that where procedure descriptions are developed for inclusion in the functional specifications, decision tables be used, for two reasons.

1. They are easier to read than narrative descriptions.
2. They force consideration of possibilities that tend to be glossed over in narrative descriptions.

The quality of the procedure description is enhanced if a comprehensive set of test case examples is included. By the time system development begins to concentrate on design activities, considerations begin to get more technical and less functional. As a consequence, the user's grasp of what's being specified may begin to suffer. Nevertheless, at the end of system construction it's the user who must be satisfied. The user may get lost in the technicalities of the design specification, but he should understand the data that's being processed. Consequently he's in a position to pass on the accuracy of the test case examples developed, or even better, to develop them himself. A comprehensive, user-prepared set of test data examples is a long step toward a satisfied user.

For example, suppose the discount policy is that no discount is given on invoices totaling $1000 or less, a 10% discount is given on the amount of the invoice that is more than $1000 but less than or equal to $5000, and a 20% discount is given on amounts in excess of $5000. Then the following examples clear up any confusion as to what the discount policy is.

Example 1

Invoice Gross Amount			$1,000
Invoice Net Amount			$1,000

Example 2

Invoice Gross Amount			$5,000
Invoice Net Amount		$1,000	
+ 0.90 ($4,000)	=	3,600	$4,600

Example 3

Invoice Gross Amount			$10,000
Invoice Net Amount		$1,000	
+ 0.90 ($4,000)	=	3,600	
+ 0.80 ($5,000)	=	4,000	$ 8,600

Any performance characteristics of the system that are pertinent to the user are also specified in the processing section. For example, if the system is realtime, standards for response and uptime are specified.

A well executed procedure description is generally a good-sized document. One of the most effective ways you can make such a document readable is to design it the way an onion is constructed, so it can be peeled away a layer at a time. This means the first thing in the description is a relatively short, general description of the procedure that gives an overall picture of what's going on. You follow this general description with the detailed description. If the detailed description is large, you divide it into sections. You organize each section into a general description of the information in the section, followed by the detailed description. If the detailed description of a section is large, you divide it into subsections, each consisting of a general description and a detailed description, until the detail in any one subsection becomes small enough to be easily readable.

What is "small enough" isn't best defined in terms of number of pages. Information is organized into functional units, and the breaking of a functional unit into two subsections to conform to some maximum page standard doesn't contribute to readability. A better rule is to divide the document into as many levels of subsections as organization of the content allows.

All this leads to the conclusion that a good step in the preparation of the organization of your procedure description is to construct a hierarchy chart of the functions making up the procedure. Such an approach has the secondary benefit of laying the groundwork for a top-down design. Figure 6.1 is an example of a hierarchy chart.

6.3.1.5. Control. This section of the system description spells out the audit procedures, control techniques, and provisions for security and privacy the user wishes to have followed.

6.3.1.6. Some Comments. I've been exposed to few good examples of functional procedure descriptions in the application world, and I've been exposed to lots of bad ones. The good examples I've seen tend to describe conversational systems. Data processing personnel seem to realize that when they're developing a conversational system, they'd better get the user-system interface spelled out in considerable detail and get it approved by the user early in the game if they expect any success in the system development effort. They don't seem to realize that the input and output of a batch system represent the same interface, and its definition is often ignored to the demonstrable detriment of the system developed.

I know of two areas where a good job of developing functional procedure descriptions is done. One is in engineering. When a project to develop a new computer system is established, one of the first steps is to settle on a set of functional specifications. This document bears a close resemblance to the principles of operation manual. It contains no reference to logic design or components, and it's agreed on before significant work on logic design begins.

One segment of the data processing world that seems to have learned its lesson in the functional procedure description area is systems programmers. When a group is organized to develop an

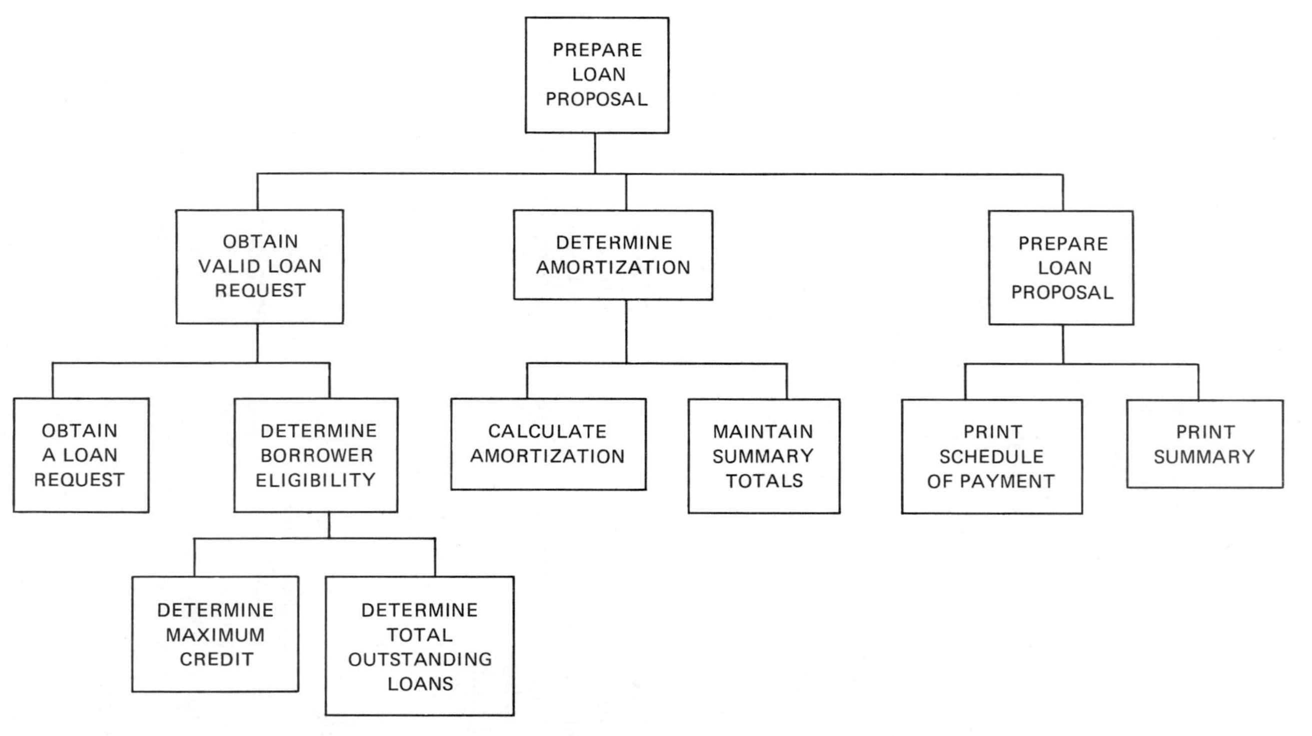

Figure 6.1 Hierarchy chart.

assembler, for example, the first thing they prepare is a set of functional specifications describing what the assembler will do. With minor modification this document ultimately becomes the assembler user manual. Systems programming functional specifications aren't perfect, but applications system development people would experience a great leap forward in productivity if they did as well.

Finally, we should note that user comment on the functional adequacy of a system most often surfaces when the user begins to see the first output from the system. This point occurs late in system development, a situation which has led to the sardonic comment that the way to build a system right is to build it twice—once to get the user's real input, and then a second time to get it right. In response to the flash of insight in such a remark, we can say that the more the user is actually put into the situation of seeing the system he's specifying at the time the functional specifications are being developed, the more reliable the specifications will be. All this argues for the use of models, automated or clerical, to simulate system operation for the user's review.

6.3.2. Acceptance Criteria

This section describes how the user will be satisfied that the completed system operates properly. Such a demonstration calls for establishing criteria for measuring system accuracy. The user must prepare the inputs in full volume for, and review the outputs of, the acceptance demonstration. This section specifies who will do this job and how.

If the new system is replacing an old one, either manual or automated, then it may be possible to conduct a parallel test. For example, the user may specify that his people will compare the results produced by the new method with those produced by the old, and after three successive cycles of successful comparison (as "successful" is defined) he will be satisfied of the accuracy of the new system.

In some cases a parallel test may not be possible. Here the user may stipulate, for example, that his people will randomly select 5 percent of all transactions submitted, manually process them, and compare the results with the output of the new system. The user should always specify an accuracy range that's acceptable.

6.3.3. Measuring Benefits

After a computer system is installed and operating, the question "Was it worth all that money to develop?" is appropriate. Often this question goes unanswered. The best way to prevent this from happening is to determine how the claimed benefits are to be measured. But why, you may be thinking, should this be done so early in system development? There are at least two reasons.

1. It may be that, by having to decide how a claimed benefit is to be measured later, it becomes clearer what the nature and value of the benefit is. This is the time to make this clarification, since a new feasibility decision follows completion of the specifications phase.
2. Frequently, it's easy to build into the system the means of measuring a benefit. For example, if your system is to reduce inventory levels by 10 percent, why not start accumulating in the system's historical records the information required to be able to evaluate inventory levels at any point in time, according to the criteria established in the cost benefit analysis?

EXERCISES

1. What is the product of the functional specification activity?
2. What are the three parts of the functional specifications?
3. What are the five parts of the system description section of the functional specifications?
4. What is described in the output section of the system description?
5. What is described in the input section?
6. To what extent is the description in the processing section carried out?
7. Does the processing section need to be written from scratch? Why?
8. What makes the processing section most meaningful to the user?
9. What does the acceptance criteria section specify?
10. What does the measuring benefits section specify?
11. A number of statements follow. For each one, decide whether its inclusion in a set of functional specifications would be appropriate.

    a. This information is obtained with an on-line terminal.
    b. Punch a card in the branch office location; then mail it in.
    c. The vendor number, name, and address are obtained from the permanent records.
    d. The new sales order file is input to the combined edit-sort run.
    e. The overtime pay rate is one and one half times the regular pay rate.
    f. The file is stored on disk.
    g. The system will run on the equipment currently installed in the computer center.
    h. The monthly sales analysis report must be on the sales manager's desk on the third working day of the month.
    i. Ninety percent of the messages must be responded to within three seconds.
    j. Indexed sequential file organization will be used.

7

Form Design

Each of the thirteen major sections of this chapter deals with an aspect of form design.

1. Forms can be classified into types along a number of dimensions. These classifications, which clarify some form design considerations, are made in Section 7.1.
2. As form designers, we're interested in all aspects of form design. However, form users are concerned only with form content. The extent to which form users aren't distracted from form content by other considerations of form design is a measure of our success as form designers. Section 7.2 discusses form content.
3. Ease of form use is directly related to the way the content is laid out on the form. Section 7.3 is concerned with form layout and how it can be used to direct the attention of the form user to the content.
4. Completed forms are frequently maintained in files for various periods of time. Section 7.4 discusses the impact of these filing considerations on form design.
5. Section 7.5 deals with the name to give a form.

6. Section 7.6 addresses itself to the additional form design considerations required by use of multipart forms.
7. With the introduction of computers into data processing, a large number of reports are now created on the computer's high-speed printer. Section 7.7 is concerned with how the use of this equipment influences form design.
8. Section 7.8 discusses the selection of paper stock for forms.
9. Forms carry information in two ways: encoded and in the clear. People are more at home with information in the clear, but encoded information is often more amenable to mechanical handling. Thus a problem frequently arises where it's desirable to collect information from people who think in the clear and transform it to an encoded form for mechanical handling. One solution to this problem is to encode the information automatically as it's being transcribed in the clear by people. Section 7.9 describes several ways such automatic encoding can be done.
10. Information to be processed mechanically must be in a machinable form. But it's most frequently collected on forms in handwritten or typewritten form. As a consequence, when information is to be processed mechanically, it's a common practice to go through a transcription step, during which the handwritten or typewritten information is transcribed into mechanical form. Section 7.10 discusses some techniques for avoiding this transcription step by automatically generating machine-coded documents at the time they're being completed by human information suppliers.
11. Section 7.11 directs itself to the form design considerations created by a special form of automatically generated machine-coded document, the scannable form.
12. Form design is an iterative process that strives for an optimum compromise between often conflicting considerations. The form must then be tested and perhaps modified. Even after the form has been put into use, possibilities for improvement will appear. Section 7.12 discusses various aspects of form improvement.
13. A common problem is the proliferation of overlapping and duplicate forms. The solution to this expensive problem is forms control, with which Section 7.13 is concerned.

7.1. FORM TYPES

Some forms, such as purchase orders, applications, requisitions, shop orders, and claims, are used to initiate action. These can be called *action forms.*

Other forms, such as policy files, check registers, and inventory records, are devices used to retain useful information. Such forms are called *records.*

Still other forms, such as status reports, monthly sales summaries, and market positions, are used to bring pertinent information to the attention of appropriate people. These forms are called *reports.*

Any form can be classified as an action form, record, or report form, or some combination thereof.

Information recorded on forms constitutes *data,* and forms can also be classified with regard to their function in a data processing system. Data processing systems are usually organized around the data they maintain. For example, a payroll data processing system maintains payroll records, an inventory control data processing system maintains inventory records, and so on.

In a data processing system, these records are organized into files. Thus there may be a file of payroll records, a file of inventory records, a file of insurance policy records, and so on. It's in these files that information with a relatively long-term life is retained. Consequently, from a data processing system point of view, all record forms can be considered an integral part of the system.

A data processing system may be completely manual, totally automated, or some compromise between these extremes. As a consequence, record files can appear in a number of physical shapes: as a series of forms or ledger cards in any one of a variety of file cabinets, as a deck of punched cards, or as information recorded on magnetic tape or disks.

A data processing system communicates with its environment primarily through the use of forms. Forms that provide information from the environment to the system are called *input forms;* those that communicate from the system to the environment are called *output forms.*

Many input forms are also action forms. They serve a dual purpose.

1. To initiate action in the environment
2. To tell the system the action is being taken so the system can update its records

For example, a requisition for withdrawal of material from inventory might be done by submitting a properly completed form to inventory. The form would also enter the data processing system so the inventory records can be maintained.

Another common type of input to a data processing system is a request for information being maintained in the system's files. For example, a bank officer may request information about an account balance before approving a large withdrawal.

Much of the output of a data processing system is reports. However, data processing systems also produce action forms, such as reorders to replenish stock.

A data processing system may also produce records, such as a listing of transactions processed, which are filed for some period of time. Such records are generally of a backup nature and are referred to only in case it becomes necessary to re-create or correct other data records maintained within the system.

Forms may also be classified on the basis of whether they are *single-part* or *multipart forms.*

7.2. FORM CONTENT

Designing a form is no different from designing any other part of a data processing system: the designer is responsible for the design of the form, but ultimate responsibility for determining the content of the form must lie with the form user. What is and isn't contained in the form must be worked out in cooperation with all parties concerned and must be approved by all these users. Common failures in this area are to design a form that:

1. Fails to include all information needed to fulfill the form's purpose
2. Calls for superfluous information
3. Is produced in an insufficient number of copies

However, despite the dominance of the user's role in determining form content, the designer can't be passive during this stage of form design. There are several ways the designer can aid the user in determining form content.

1. When designing an input form, the designer must continuously keep in mind that the purpose of the form is to introduce *new* information into the data processing system. Using an input form to collect information that the system is capable of generating is to make the information on the form redundant and to complicate unnecessarily the task of completing the form. For example, if in a payroll system an employee is identified on an input form by an employee number, then it's redundant to identify the employee by social security number as well, even if the system requires both identifications. The system

can be designed to interrogate its own files to determine the appropriate social security number.

2. Provision for the entry of certain information on a form may defeat the purpose of the form or have other undesirable consequences. For example, even if discounts are given to certain preferred customers, provision on a standard invoice form for entry of a discount may increase unwanted requests for a discount.

7.3. FORM LAYOUT

There are three major factors to be considered when determining form layout.

1. The ease with which the form can be completed, if it is to be completed by a person
2. The ease with which the form can be used after it has been completed
3. For printed forms, the restrictions imposed by the printing process

Restrictions imposed by the printing process (multilith, offset, etc.) must be paramount in determining form layout. However, these broad restrictions still allow considerable latitude in form layout. In the case of forms completed on a high-speed printer, the prime consideration in determining form layout is how easily it can be used after it has been completed. However, for forms that must be completed manually, ease of completion must be given priority over ease of use in determining form layout.

These factors are discussed in more detail in the next three sections.

7.3.1. Ease of Completion

Factors to consider in making the form easy to complete are as follows.

1. People who fill out forms become discouraged if they're required to provide the same information more than once.
2. Providing information can be facilitated by the sequence in which it's recorded on the form.
3. To obtain good information with a form, the form must provide adequate space in which to record the information.

4. Use of a typewriter or other mechanical device for completing a form creates special form design considerations.
5. It's sometimes necessary to ask that the same information be entered on a number of forms. There are ways to minimize the offensiveness of this requirement.
6. To be completed properly, a form must provide instructions on how to complete it.

We will discuss each of these factors separately and then give examples of both good and bad form design.

7.3.1.1. Handling Repetitive Information. The need to provide the same information more than once is often caused by having a number of similar forms each of which serves a different but related purpose. The best solution to this problem is to combine the related forms into one consolidated form that requires the entry of each unit of information only once and is then used for multiple purposes. An example is a shipping order and an invoice. Design of such a form may require the development of a multipart form, but the increase in the ease with which the form can be completed justifies the step.

Much repetitive information required on a form doesn't need to be provided by the person completing the form; instead, it can be preprinted. Dollar signs are an example of such information. Use of this technique is particularly helpful when handwriting is to be used to complete the form.

7.3.1.2. Organization of Information. Information should be organized on a form to:

1. Minimize the amount of work a person must do to complete the form
2. Allow the information to be supplied in a logical sequence

The simplest form to complete is one that requires only a checkmark in the proper place. Multiple choice questions are the best type for such a form. Such questions also tend to avoid inappropriate, ambiguous, and incomplete answers. For example, the form shown in Figure 7.1 is preferable to "List all the diseases you've had."

CHECK ALL THE DISEASES YOU HAVE HAD

☐ CHOLERA	☐ KIDNEY
☐ DIPHTHERIA	☐ MEASLES
☐ HEART	☐ PNEUMONIA

Figure 7.1

Another advantage of multiple choice format is that information is always presented in the same order. Subsequent use of the completed form is thus simplified.

Much information has a logical order, or at least a familiar one. For example, a series of calculations has a logical flow, and the common sequence for an address is street number, street name, city, state, and zip code. Since we normally read and write from top to bottom and from left to right, a request for information should be organized on a form in its logical or familiar sequence from top to bottom and from left to right. Thus an income tax form is designed so that information is requested in the order in which it's developed. Similarly, you may be requested to give your name in first, middle, and last or in last, first, and middle order, but never in middle, last, and first order.

On some forms it's necessary to provide space for optional information. The more frequently information must be supplied, the more to the top and to the left in the form the request for it should appear. For example, in the partial form shown in Figure 7.2,

1. Invoice number, requisition number, and shipping date are always required.
2. Billing number and contract number are required 60 percent of the time.
3. Serial number is required in half of the cases in which billing number and contract number are required.

INVOICE DATE	REQ. NO.	SHIP. DATE	B/L NO.	CONTRACT	SERIAL NO.

Figure 7.2

The advantage of this format is that the person completing the form works across the line only as far as he has information to supply. It's not necessary to enter information on the left and then jump across several fields to enter information on the right.

Information to appear on a form often clusters into related groups. For example, the typical purchase order contains the following types of information.

1. Information relating to the items to be purchased. For each item, the part number, description, quantity, unit price, and extension of the price must be supplied.

2. Summary information, such as the total price, any appropriate discounts, sales taxes, shipping costs, and so on
3. Vendor information, such as name and address
4. Shipping instructions: transportation method, routing, crating, etc.
5. The authorization: signatures and dates

The form is easier to complete if related information is grouped, or *zoned,* on the form, as suggested by Figure 7.3. Zoning particularly simplifies form completion when different people must complete different parts of the form.

The zoning of a form is made more apparent if the zones are set off from each other by lines or some other method.

The zoning of a form isn't necessarily done all at one level. A form can be divided into several major zones, and each of these zones can be further divided into subzones, and so on. Lines of different widths can be used to signify the level of the zone—the heavier the line, the more major the zone. With a multipage form, each page can represent a major zone of information. The federal income tax form is zoned this way.

On forms on which a person may not be required to fill out all

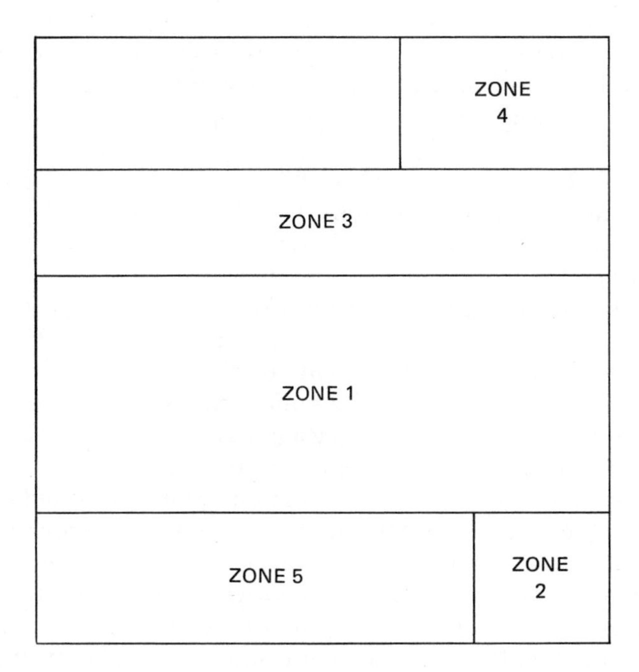

Figure 7.3

entries, there may be certain entries that are mandatory. Attention can be drawn to these entries in several ways.

1. Place them near the top and left of the form.
2. Separate them from other entries by a heavy or colored line.
3. Use larger type than that used on the rest of the form.
4. Use attention getters, such as arrows, pointing fingers, or colored ink.

7.3.1.3. Space Requirements. A form must allow enough space so a person completing it has no difficulty entering the information requested. On the other hand, excess space gives the impression that more detail is desired than is actually the case. The only way to tell whether appropriate space has been provided is to test-use a proposed form under a wide variety of conditions.

Be sure to leave enough space for all entries on a form. For example, if information is to be entered on a form in the clear and is then to be encoded before being keypunched, be sure space on the form is provided both for the information in the clear and for the entry of the appropriate codes.

In some instances, the entry of all requested information requires a multipage form. One way to design a two-page form is to print the form on both sides of a sheet of paper. If you make such a choice, be sure the person completing the form realizes there's information to be supplied on both sides. Place the statement "Please complete *both* sides of this form" prominently at the top of the front of the form.

7.3.1.4. Using a Typewriter. When possible, avoid the use of handwriting for completing forms. It's often illegible.

If a typewriter or other transcribing machine is used to fill out a form, design the form for ease of machine use.

The horizontal lines of the form should conform to the spacing provided by the carriage-return lever. This eliminates the need for the typist to make fine adjustments to align the type with each line. Horizontal lines should have the same rachet spacing throughout the form. Make the form all single-spaced or all double-spaced.

To the extent possible, the left-hand and right-hand margins of all forms used by any one installation should be standardized to eliminate the need to reset margin controls each time a different form is prepared. Standardization of tabulation stops should also be attempted. At the least, tabulation stops on any given form should be standardized so there's a minimum of tabulating to complete the form. Box design facilitates standardization of tabulation stops. (Box design is discussed in Section 7.3.1.6.)

Design the form so it can be filled in by typing directly across the form. Don't design a form that requires the typist to back up the platen. Minimize the number of carriage returns and skips required. Box design helps reduce the number of skips required.

Somewhat in contrast to the rule when the form is to be completed by handwriting, a form designed for use with a typewriter shouldn't have all possible information preprinted on it. For example, don't separate dollars and cents with a line or a preprinted decimal point. Let the typist type the decimal point. Such an approach makes it easier for the typist to align the form when it's inserted in the typewriter, and the completed form is neater and more legible.

Both the vertical and horizontal layout of the form must conform to the characteristics of the machine on which it's to be completed.

1. Generally, there are six horizontal lines to the inch if the form is to be single-spaced, three lines to the inch if double-spaced.
2. Specially designed graph paper can be used to design typewriter forms. The Hammermill Paper Company makes form layout sheets for both pica and elite typewriter spacing.
3. An "all-purpose" spacing can be used. All-purpose spacing allows three horizontal lines to an inch and five characters to an inch along a line. Such spacing is compatible with pica type, elite type, and handwriting.

But the indispensible way to determine whether the form has been laid out for easiest use on the equipment for which it's designed is to try it out on the equipment after you've laid it out.

7.3.1.5. Entry of the Same Information on More than One Form. If more than one item of information must be entered on more than one form, the information to be supplied should be laid out in the same way on each form. Such an approach is called *interlocking* the forms. The advantages of interlocked forms are that:

1. The sequence in which the information is entered becomes fixed in the mind of the person completing the forms, and this makes it less difficult to supply the information repeatedly.
2. If the information must be transcribed from one form to another or if the information on two forms must be compared, interlocking the forms makes the transcription or comparison easier and reduces errors.

7.3.1.6. Instructions. Instructions for filling out a form should be clear, concise, complete, and useful.

Use simple, grammatical, properly punctuated language. Try not to use trade jargon; if you must, be sure the terms are in current use. Try to avoid abbreviations; if you must use them, follow standard usage. Table 7.1 gives a sample list of standard abbreviations.

Table 7.1

Word	*Caption*	*Word*	*Caption*
Accumulative	Accum	Miscellaneous	Misc
Account	A/C	Model	Mod
Amount	Amt	Month	Mo
Assembly	Asmb'ly	Number	No
Authorized by	Auth by	Operator	Oper
Average	Avg	Operation	Oper'n
Balance	Bal	Paid	Pd
Check	Ck	Paragraph	Par
City	Cty	Pieces	Pcs
Department	Dept	Production	Prod
Discount	Disc	Purchase Order	P O
Drawing	Drw'g	Quantity	Quan
Each	Ea	Received	Rec'd
Estimate	Est	Reference	Ref
Freight	Frt	Rejection	Rej'n
Hours	Hrs	Required	Req'd
Including	Incl'g	Requisition	Req
Inspect	Insp	Revision	Rev
Insurance	Ins	Rework	Rwk
Invoice	Inv	Schedule	Sched
License	Lic	Signature	Sig
Machine	Mach	Superintendent	Supt
Manager	Mgr	Symbol	Sym
Manufacturing	Mfg	Weight	Wt
Merchandise	Mdse	Year	Yr

Instructions are most effective when included directly on the form. One way to do this is by means of *captions.* A caption labels the area in which information is to be entered and identifies the information to be entered.

The best way to use captions is to use *box design,* that is, to provide an outlined space (a box) in which to enter the information.

The caption is placed in the upper lefthand corner of the box (see Figure 7.4). Box design maximizes the space on the form for entry of information and puts the captions in a place where they don't distract the person using the completed form.

Captions should be printed in an easily readable type face. However, the person using a completed form doesn't want to be distracted by the captions, so the captions shouldn't be printed in type any larger or heavier than is required by considerations of legibility. Captions can also be printed in a recessive color, such as green, blue, or brown, so the information entered on the form will stand out.

Captions must be complete. If what you want entered in a box is the assembly part number, don't caption the box PART NUMBER, caption it ASSEMBLY PART NUMBER. If what you want entered is the date an order is shipped, don't caption the box DATE, caption it DATE SHIPPED.

If you want information entered in a box in a particular way, spell the format out in the caption. (For example, see Figure 7.5.) A box simply captioned DATE doesn't say whether "month, day, year" or "year, month, day" is desired, nor does it indicate whether the month is to be spelled out or shown as a number. The form shown in Figure 7.6, which indicates both, is preferable.

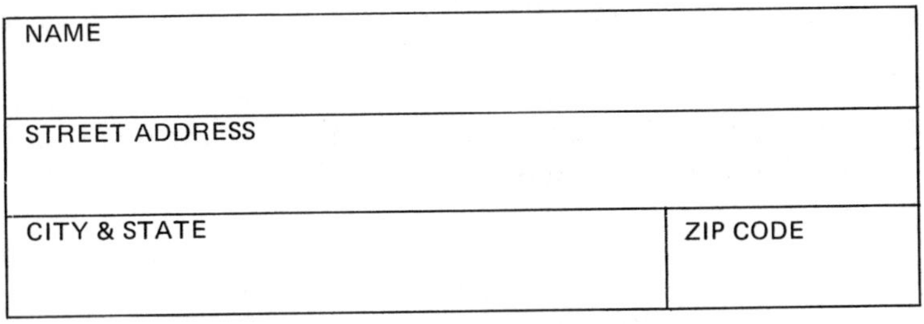

NAME	
STREET ADDRESS	
CITY & STATE	ZIP CODE

Figure 7.4

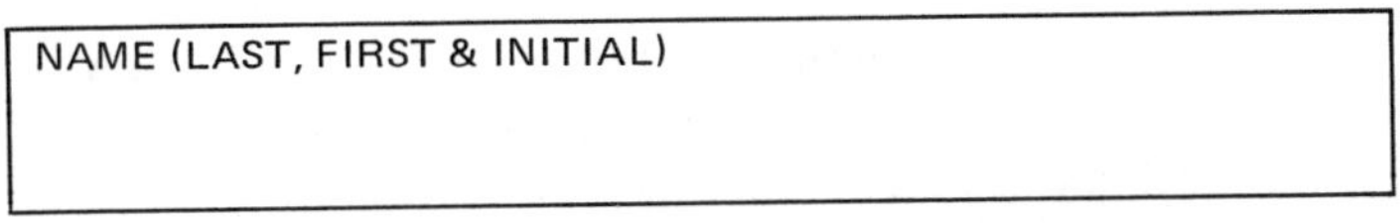

NAME (LAST, FIRST & INITIAL)

Figure 7.5

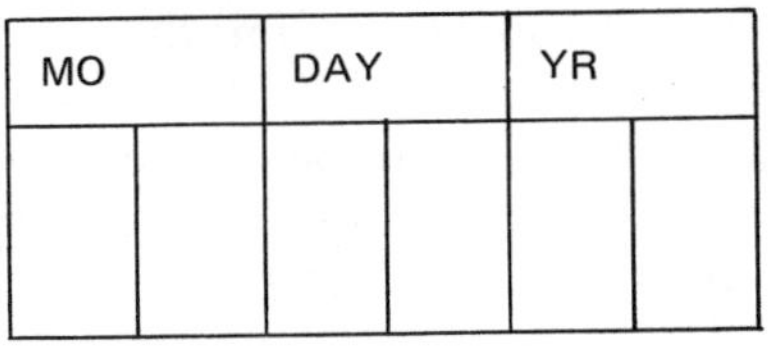

MO		DAY		YR	

Figure 7.6

In addition to captions, other instructions, both for filling out the form and for routing the form after it has been completed, can be incorporated into the form. One way to draw attention to such instructions is to print the form in one color ink (usually black) and the instructions in another (usually red).

Instructions can be printed on the back of the form. However, if such an approach is taken, a notice to the effect that they're on the back must be printed prominently on the front of the form. Otherwise, people may complete the form without realizing that instructions are available.

However, flipping the form back and forth, reading instructions and entering information, is awkward. An alternative is a hinged and perforated paper that, when opened, has the instructions facing the form. Since more people are right-handed than left-handed, put the instructions on the left so the writing arm won't obscure them. After the form has been completed, the instruction page can be torn off, and the completed form alone can be routed for further processing. If forms are supplied in a pad, another approach is to print the instructions on the inside of the cover of the pad.

For long, complicated forms like the Federal income tax form, a separate instruction booklet can be provided. An advantage of an instruction booklet is that it can be used for completion of many forms.

In case the instructions are physically separated from the boxes in which the information is to be entered, number the boxes and key the instructions to the numbers.

7.3.1.7. Example. The form shown in Figure 7.7 is an example of poor layout. Note all the errors you can discover in this form before checking the list below.

1. The entries for "OUR ORDER NO.," "SHIP VIA," "INV. DATE," "CUST. ORD. NO.," and "CUST. REQ. NO." aren't standardized for tab stops.
2. These entries aren't standardized for platen ratchet spacing. Some require single-spacing, others double-spacing.
3. The entries for "INV. NO." and "TERMS" can be preprinted.
4. There's waste space on the top, bottom, and right of the form.
5. The information to be entered is scattered over the form.
6. The typist must tab to type the "SOLD TO" address.

7. The typist must back up to type the "SHIP TO" address after typing the "SOLD TO" address.
8. "QUAN. ORD." and "DESCRIPTION" are filled out at one typing; "QUAN. SHIPPED," "PRICE," and "AMOUNT" at another. The column order requires skips in both operations.
9. The "DESCRIPTION" column isn't wide enough.
10. The form requires all shipped quantities to be recorded. A more desirable exception procedure would be to record only backorder quantities.
11. Dollars and cents are separated by a line.

Figure 7.8 shows the same form, properly designed.

7.3.2. Ease of Use

Particularly in the case of reports, information should be organized on a form so the user of the completed form can readily find the information wanted.

In some instances when information on a form is to be keypunched, the sequence of fields on the card can be made to follow the sequence on the form. In such cases the keypunching can be done directly from the completed form; the form should be laid out so there's a box for each character to be entered on the form, and each box should be labeled with the number of the column in which the character is to be punched. An example of such a form is shown in Figure 7.9.

7.3.3. Printing Considerations

Usually printing can't be done on the edge of a sheet of paper. Consequently, don't extend lines or type all the way to the edge. Some printing processes require that each sheet of blank paper be gripped by the press. In such instances a margin of at least 1/4 inch (1/2 inch is better) must be left at the top or the bottom of the form.

Serially numbered forms are expensive and thus should be avoided if possible. If you must have a serially numbered form, leave enough space around the number so two printer runs aren't needed to print the form.

To be sure you've provided for all printing considerations in your form layout, you should consult with the person who will handle the printing before you complete your layout.

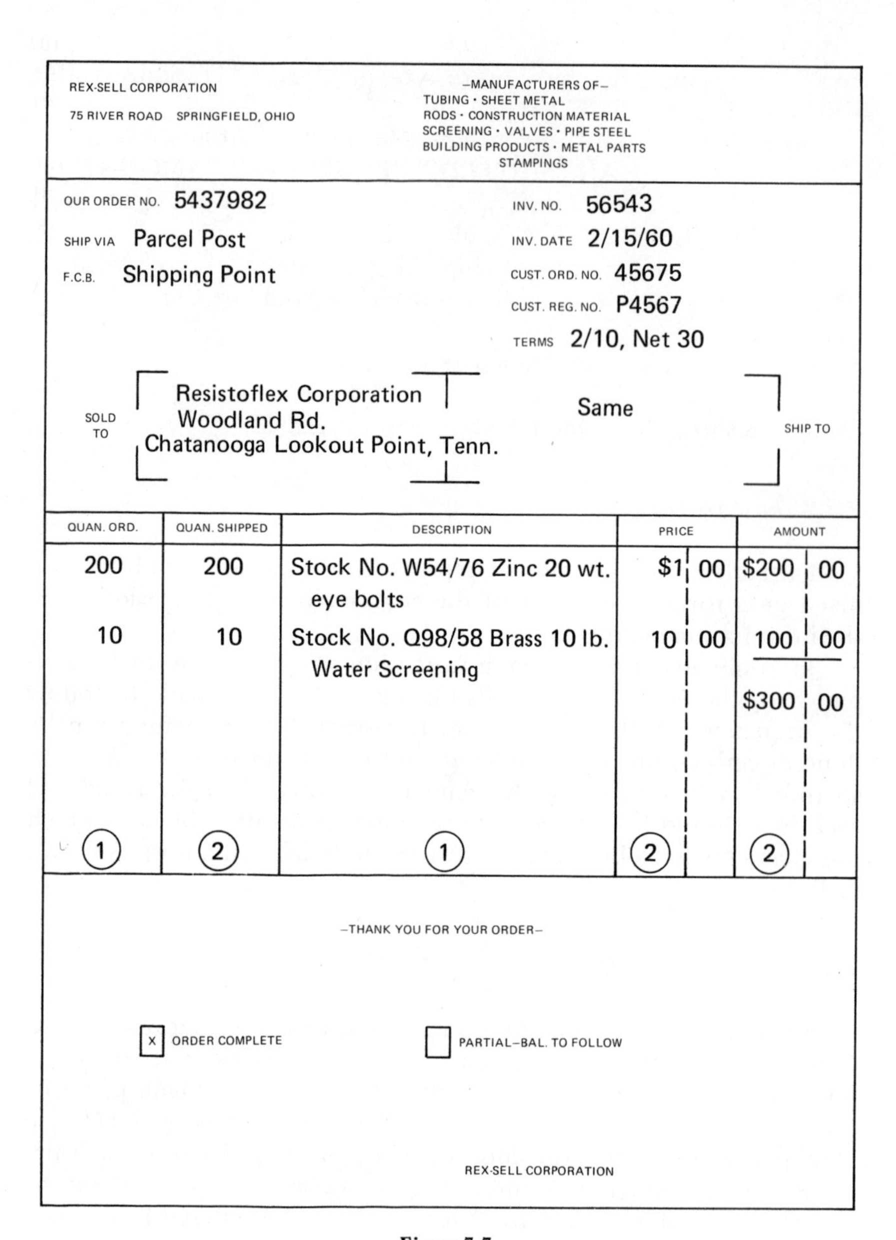

REX-SELL CORPORATION

75 RIVER ROAD SPRINGFIELD, OHIO

–MANUFACTURERS OF–
TUBING · SHEET METAL
RODS · CONSTRUCTION MATERIAL
SCREENING · VALVES · PIPE STEEL
BUILDING PRODUCTS · METAL PARTS
STAMPINGS

OUR ORDER NO. 5437982
SHIP VIA Parcel Post
F.C.B. Shipping Point

INV. NO. 56543
INV. DATE 2/15/60
CUST. ORD. NO. 45675
CUST. REG. NO. P4567
TERMS 2/10, Net 30

SOLD TO
Resistoflex Corporation
Woodland Rd.
Chatanooga Lookout Point, Tenn.

SHIP TO
Same

QUAN. ORD.	QUAN. SHIPPED	DESCRIPTION	PRICE	AMOUNT
200	200	Stock No. W54/76 Zinc 20 wt. eye bolts	$1 00	$200 00
10	10	Stock No. Q98/58 Brass 10 lb. Water Screening	10 00	100 00
				$300 00
(1)	(2)	(1)	(2)	(2)

–THANK YOU FOR YOUR ORDER–

[X] ORDER COMPLETE

[ ] PARTIAL–BAL. TO FOLLOW

REX-SELL CORPORATION

Figure 7.7

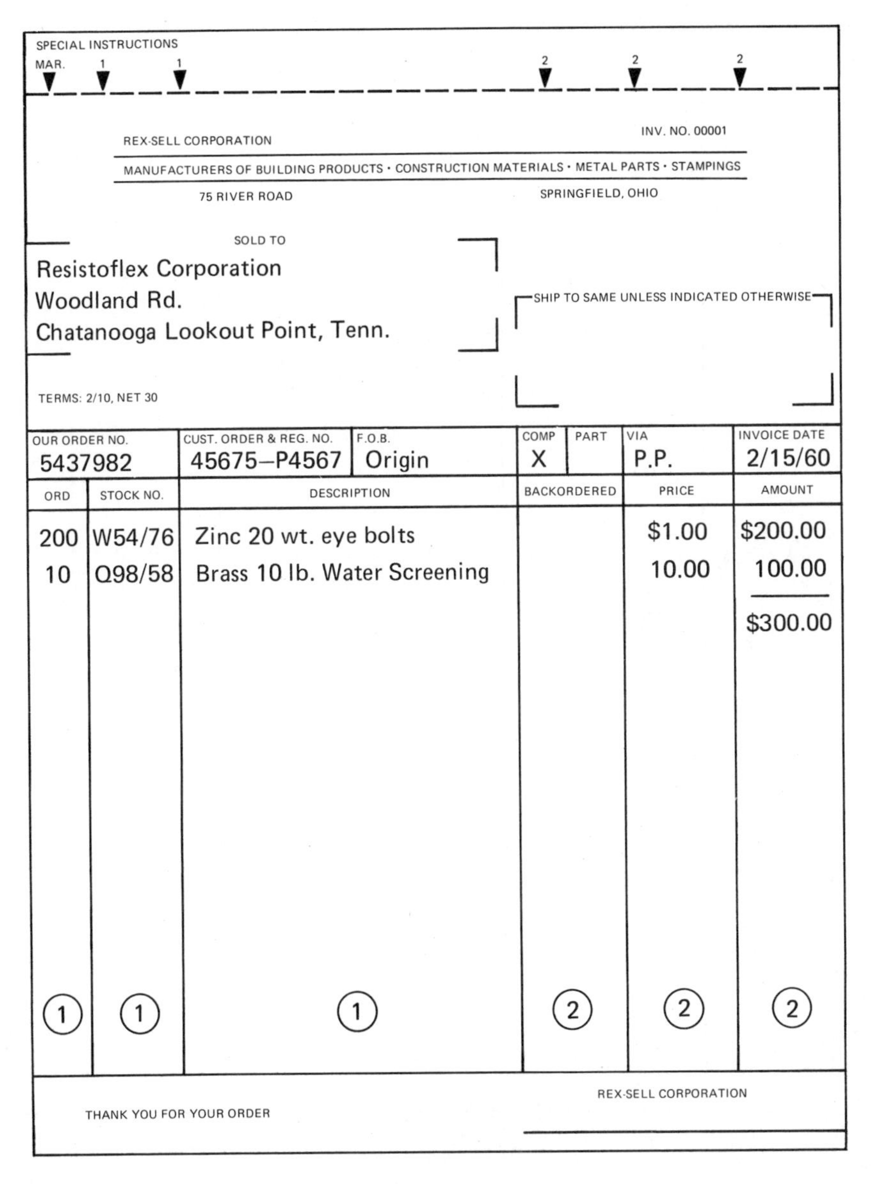

SPECIAL INSTRUCTIONS
MAR. 1 1 2 2 2

REX-SELL CORPORATION — INV. NO. 00001

MANUFACTURERS OF BUILDING PRODUCTS · CONSTRUCTION MATERIALS · METAL PARTS · STAMPINGS

75 RIVER ROAD — SPRINGFIELD, OHIO

SOLD TO

Resistoflex Corporation
Woodland Rd.
Chatanooga Lookout Point, Tenn.

SHIP TO SAME UNLESS INDICATED OTHERWISE

TERMS: 2/10, NET 30

OUR ORDER NO.	CUST. ORDER & REG. NO.	F.O.B.	COMP	PART	VIA	INVOICE DATE
5437982	45675–P4567	Origin	X		P.P.	2/15/60

ORD	STOCK NO.	DESCRIPTION	BACKORDERED	PRICE	AMOUNT
200	W54/76	Zinc 20 wt. eye bolts		$1.00	$200.00
10	Q98/58	Brass 10 lb. Water Screening		10.00	100.00
					$300.00
(1)	(1)	(1)	(2)	(2)	(2)

THANK YOU FOR YOUR ORDER — REX-SELL CORPORATION

Figure 7.8

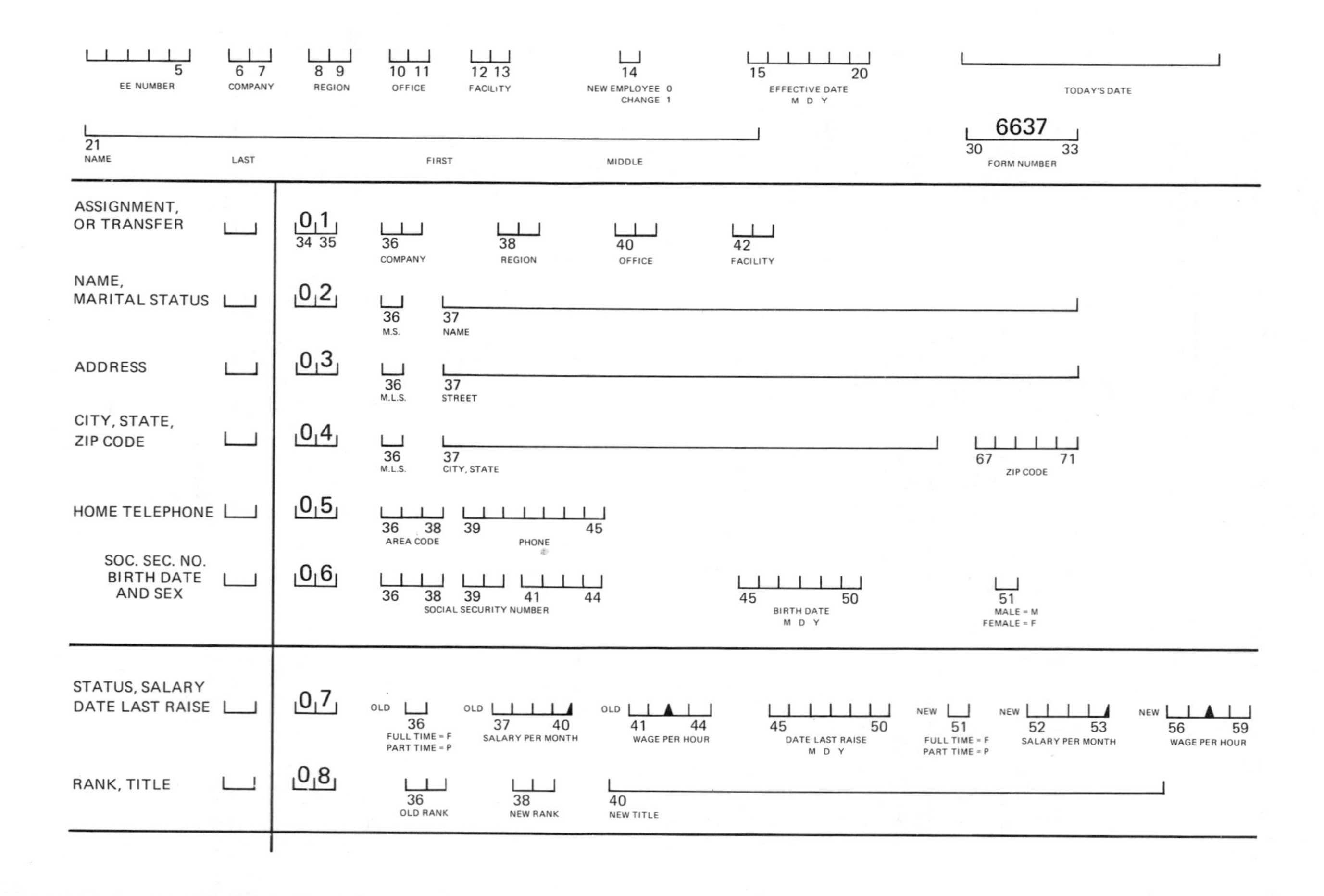

5 EE NUMBER | 6 7 COMPANY | 8 9 REGION | 10 11 OFFICE | 12 13 FACILITY | 14 NEW EMPLOYEE 0 CHANGE 1 | 15 20 EFFECTIVE DATE M D Y | TODAY'S DATE

21 NAME LAST FIRST MIDDLE | 6637 30 33 FORM NUMBER

ASSIGNMENT, OR TRANSFER — 0 1 (34 35) — 36 COMPANY | 38 REGION | 40 OFFICE | 42 FACILITY

NAME, MARITAL STATUS — 0 2 — 36 M.S. | 37 NAME

ADDRESS — 0 3 — 36 M.L.S. | 37 STREET

CITY, STATE, ZIP CODE — 0 4 — 36 M.L.S. | 37 CITY, STATE | 67 71 ZIP CODE

HOME TELEPHONE — 0 5 — 36 38 AREA CODE | 39 45 PHONE

SOC. SEC. NO. BIRTH DATE AND SEX — 0 6 — 36 38 39 41 44 SOCIAL SECURITY NUMBER | 45 50 BIRTH DATE M D Y | 51 MALE = M FEMALE = F

STATUS, SALARY DATE LAST RAISE — 0 7 — OLD 36 FULL TIME = F PART TIME = P | OLD 37 40 SALARY PER MONTH | OLD 41 44 WAGE PER HOUR | 45 50 DATE LAST RAISE M D Y | NEW 51 FULL TIME = F PART TIME = P | NEW 52 53 SALARY PER MONTH | NEW 56 59 WAGE PER HOUR

RANK, TITLE — 0 8 — 36 OLD RANK | 38 NEW RANK | 40 NEW TITLE

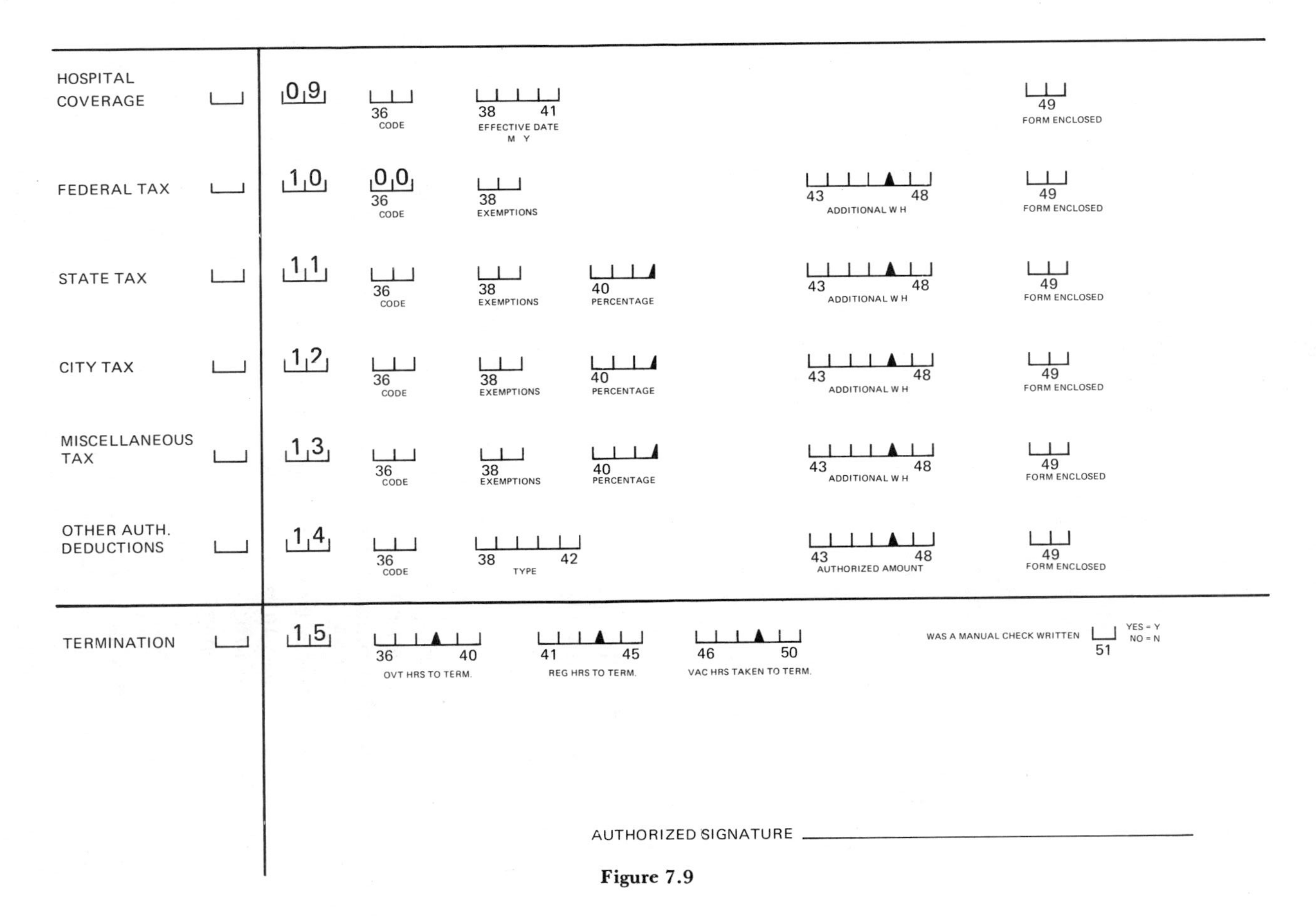

HOSPITAL COVERAGE | 0 9 | 36 CODE | 38–41 EFFECTIVE DATE M Y | 49 FORM ENCLOSED

FEDERAL TAX | 1 0 | 0 0 36 CODE | 38 EXEMPTIONS | 43–48 ADDITIONAL W H | 49 FORM ENCLOSED

STATE TAX | 1 1 | 36 CODE | 38 EXEMPTIONS | 40 PERCENTAGE | 43–48 ADDITIONAL W H | 49 FORM ENCLOSED

CITY TAX | 1 2 | 36 CODE | 38 EXEMPTIONS | 40 PERCENTAGE | 43–48 ADDITIONAL W H | 49 FORM ENCLOSED

MISCELLANEOUS TAX | 1 3 | 36 CODE | 38 EXEMPTIONS | 40 PERCENTAGE | 43–48 ADDITIONAL W H | 49 FORM ENCLOSED

OTHER AUTH. DEDUCTIONS | 1 4 | 36 CODE | 38–42 TYPE | 43–48 AUTHORIZED AMOUNT | 49 FORM ENCLOSED

TERMINATION | 1 5 | 36–40 OVT HRS TO TERM. | 41–45 REG HRS TO TERM. | 46–50 VAC HRS TAKEN TO TERM. | WAS A MANUAL CHECK WRITTEN 51 YES = Y NO = N

AUTHORIZED SIGNATURE ____________________

Figure 7.9

7.4. FILING CONSIDERATIONS

If a completed form is to be filed or bound, the file or binder to be used should determine form size. The best approach is to make all forms 8-1/2 by 11 inches in size.

If the form is to be bound, leave enough margin on the form so all the information on it can be read easily after it has been bound.

Place the identifying information on a form where it can be readily seen after the form has been filed or bound.

1. If the form is filed in a standard file cabinet or is bound on the left, put the identifying information on the right-hand side of the form.
2. If the form is bound on the top, put the identification at the bottom.

7.5. FORM NAME

Give every form a name, and put the name on the form. The name is what the form users use to identify the form. Like a caption, a form name should be meaningful and concise. It should tell what the form does.

Standardize the place where you put the name on all forms. A good place is at the top.

7.6. MULTIPART FORMS

Multipart forms are used when more than one copy of a completed form is required. With a multipart form, all the required copies are completed at once. Ways in which this is done are as follows.

1. Carbon paper can be inserted between copies of the form.
2. The forms can be procured with one-time carbon paper interleaved with the forms.
3. Carbonless paper can be used.

If the form is supplied in a pad with interleaved carbons or carbonless paper, it may be necessary to have cardboard inserted under the last copy of a set when the set is filled in.

A convenient way to package forms with interleaved carbons or carbonless paper is as a snapout set. However, this convenience must be balanced against the added cost of such forms.

Carbon paper comes in different grades. The best way to choose the carbon paper for your form is to try out samples until you get the results you want. Vendors will be pleased to offer advice as well.

Making each copy of a form a different color facilitates routing of the copies.

Multipart forms are particularly useful for a consolidated form designed to serve multiple purposes. In this case all information is entered on the original of the form, and selected parts of this information are reproduced on the copies of the form.

One way to effect this selective reproduction is to block out areas of the copies where it's not desired to reproduce the original information. If such an approach is taken, use a pattern in the area blocked out rather than a solid blacked-out area. A solid blacked-out area won't make the copy that is blocked out illegible.

However, blocked out areas create suspicion in the minds of people receiving the copy. "What is it," they think, "that *I'm* not privileged to see?" A better approach is to use interleaved carbons that are die cut to reproduce only the desired information. Zoning simplifies the layout of a form that uses die-cut carbons.

For communication with the printing company you should lay out all copies of a form that aren't replicas of the original.

If a typewriter is to be used to enter information on a multipart form, watch the vertical spacing when laying out the form. Because of the rolling of the form around the platen, this spacing will vary from one part of the form to another. The only way to be sure you've handled this problem in your form layout is to try the form out before you settle on a final layout.

7.7. HIGH-SPEED PRINTER OUTPUT

Layout of high-speed printer forms is governed primarily by ease of use of the final product. Other considerations are as follows.

You must decide whether to use blank or preprinted forms. Form suppliers make blank stock available in a variety of colors and sizes, plain or lined, and in single or multiparts. Blank stock is less expensive than preprinted forms and doesn't present the inventory problems associated with preprinted forms. The use of blank stock is often acceptable, since it's not difficult to program a computer to print repeated information such as headings, captions, and punctuation.

Here are the factors that make a preprinted form practical.

1. Regulations, such as those governing the W2 forms, may require a preprinted form.
2. If the prepared form is to be sent to someone the user wants to impress (a customer, for example), considerations of presentability may indicate the use of a preprinted form.
3. Complicated forms may have to be preprinted. For example, it may be impractical and expensive for a computer to print the company name and address and the terms of sale on every invoice.

In data processing centers, the high-speed printer tends to be a bottleneck. Consequently, anything that can be done in printer form layout to speed the printing operation contributes to the efficiency of the center. Some considerations in this area are as follows.

1. Some forms are narrow enough to allow two or more to be printed side by side at the same time. This significantly reduces printing time.
2. Keep the number of lines to be printed on a form to a minimum. Computers usually print one line at a time, and it's the number of lines rather than the number of characters to be printed that determines the time required to print a form.
3. In terms of time per line, a printer can usually space over a large number of lines faster than it can space over one or two. Consequently, try to cluster into adjacent lines the information to be printed on a form.
4. Printer time is lost whenever one stack of continuous forms must be removed from the printer and another inserted. Avoiding this setup time is an argument for standardizing on a single blank stock to be used for all output. Such a goal can be approached but isn't usually completely attainable. Ways to minimize setup time when forms must be changed are as follows.
   a. Standardize the left-hand margin of all forms. This allows the left-hand tractor of the printer to remain in the same position for all output and consequently simplifies paper alignment.
   b. Use a standard form width. Such a standard immobilizes the right-hand tractor as well.

Form alignment on a high-speed printer is always less than precise. Consequently, always design a preprinted form so that if alignment is approximately correct, the completed form is legible and neat. For example, wherever a preprinted form has a preprinted line, a carbon edge, or a perforated edge, allow at least one character space in which no printing occurs.

The number of copies a high-speed printer can produce is limited. If more copies are desired,

1. The printing operation can be repeated.
2. A reproducible master can be printed, and this master can be used to produce additional copies on a reproduction device.

The choice is economic and must be determined on the basis of the equipment available, the copies required, and the volumes to be copied.

7.8. PAPER SELECTION

The use to which a form is put determines the type of paper on which it should appear. Some pertinent questions are:

1. Is the form going to be subjected to rough use? Will it be bent and folded?
2. Will people do a significant amount of erasing on the form?
3. Is it necessary to allow for the use of ink in filling out the form? (If the paper is too porous, ink can't be used.)

A good, standard form paper is No. 4 sulphite bond. However, your best bet is to determine the use your form will receive and then consult with the person supplying the form to determine the kind, grade, and weight of paper appropriate to your form.

7.9. AUTOMATIC ENCODING OF INFORMATION

Automatic encoding is concerned with arranging for people to supply information in coded form without being aware of the coding structure. Ways to effect automatic encoding are:

1. Stamps and plates can be used. Perhaps the most common example is the credit card or charge plate. Use of such a plate removes the necessity for handwriting names and addresses, which promotes legibility. However, the most important advantage is that the plate automatically provides the customer's correct account number.

2. A form can be laid out so that, when indicating the appropriate information, one also automatically selects the code for the information (Figure 7.10).
3. A form can be organized so that, when selecting the appropriate information in the clear on the original of the form, one automatically encodes the information on a copy. (Figure 7.11).

7.10. AUTOMATIC GENERATION OF MACHINE-CODED DOCUMENTS

Automatic generation of machine-coded documents serves the purpose of automatically generating information in a form suitable for machine handling at the time the information is supplied.

One technique is to provide the person supplying the information with a spot punch (similar to a train conductor's punch) and have him punch the information into a punch card. The drawback of this approach is that it requires the person filling out the form to know punch-card code.

An improvement over this technique is to supply the person with a portable punch. Such a punch has a keyboard for numerics, but if

CHECK APPROPRIATE BOX WITH AN (X)

	04	DEPARTMENT SUPERVISOR
	16	FOREMAN
	05	ASSISTANT FOREMAN
	21	INSPECTOR
X	14	ASSISTANT INSPECTOR
	53	ASSEMBLER
	39	MACHINIST
	52	SET UP MAN
	68	STOCKMAN

Figure 7.10

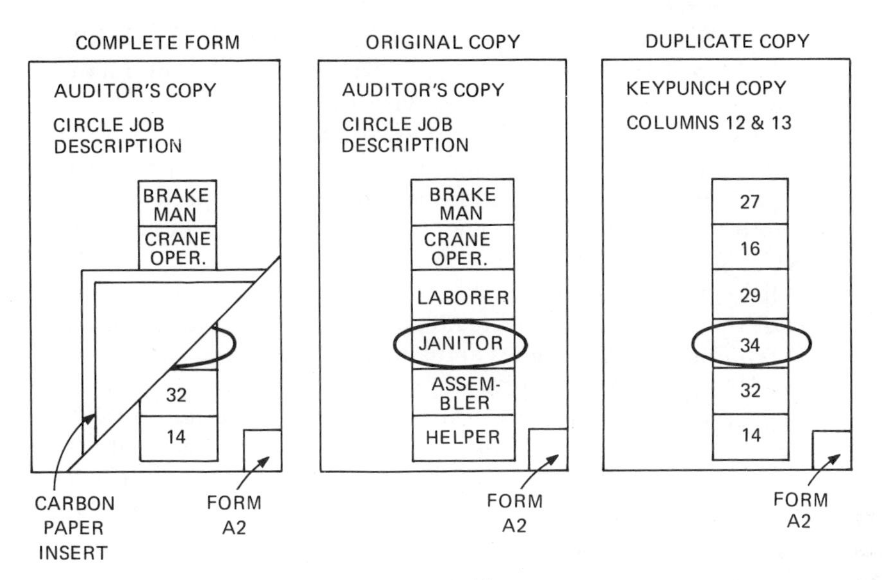

Figure 7.11

alphabetic information is required, knowledge of punch-card code once more becomes a requirement.

Use of mark-sensed cards is the next step in this progression. Here no knowledge of punch-card code is required. But because the method of completing the form is unfamiliar, this method requires special training.

The goal of these techniques is to have the person record the information on the form in a machinable manner. Another general approach is to have the machinable document produced as a byproduct of producing a typewritten or other keyboarded document. Paper tape or card punching equipment can be attached to a variety of electric office machines, such as typewriters, adding machines, bookkeeping machines, and cash registers. These punches then produce machinable documents as a byproduct of the operation of the office machine.

No matter what approach is taken, two types of information are typically being supplied.

1. Information unique to this transaction (for example, the quantity of items on a purchase order)
2. Information common to many transactions (for example, vendor name)

While the unique information must be supplied by a marking or keyboarding operation, the common information can be supplied in a more automatic manner. For example,

1. If the information is being punched into cards, a tub file of gangpunched cards can be used. Thus for each vendor a number of cards are prepunched with the vendor's name and address, and these prepunched cards are stored in a reservoir called a tub file. When an order is received, a card prepunched with the appropriate vendor name and address is selected from the tub file, and the unique order information is punched into it.
2. Office machines equipped with punches can also be equipped with paper tape or card readers. A master set of tapes or cards containing vendor names and addresses is maintained. When an order is received, the appropriate master tape or card is selected and inserted into the reader. The master causes the vendor name and address to be typed and punched in the output tape or card. The unique order information is then keyboarded, and this causes the information to be both typed and punched.

Another way to generate machine-coded documents automatically is illustrated in the following example. Suppose a computer prints bills and, for each bill printed, simultaneously punches a card containing the information printed on the bill. The card is sent to the subscriber with the bill, and the subscriber is instructed to return the card with payment. If the bill is paid as printed, the card is an automatic input document to indicate to the computer that the bill has been paid.

One problem with the generation of both a printed bill and an associated punched card is the matching of the bills with the cards after they've been produced. Sporadic attempts have been made to develop equipment that will efficiently punch a card and also print on it. Such equipment is desirable as a solution to this problem, because the punched card can also serve as the printed bill, thus eliminating the need for the matching operation.

If you can confine the printed information on a bill to what's punched in the card, a currently available solution to this problem is to have the computer simply punch the bills and then interpret the cards.

7.11. SCANNABLE DOCUMENTS

The ultimate resolution of the problem of people who record information in one way and machines that read information in another seems to be to design equipment that reads information in the way it's recorded. It's to the attainment of this goal that OCR (Optical

Character Recognition) equipment is directed. The design of this equipment is presently in a state of flux, and if you're designing scannable documents for use with such equipment, your best bet when laying out your form is to work closely with a representative of the company supplying the equipment.

7.12. FORM IMPROVEMENT

Form layout isn't the final step in form design. It's one of the first. After the form has been laid out, it must be subjected to a field test under as wide a variety of conditions as possible to see whether the layout is as effective as it can be. Chances are such a field test will suggest ways the layout can be improved.

Even after a form has been put into production, the users will continue to see ways it can be improved. Every form should be controlled from a central point in an organization, and suggestions for improvement can be maintained in a form's file at this central point. These suggestions can be incorporated the next time the form supply is replenished.

One way to recognize possibilities for improving a form in use is to inspect a number of completed forms. Many difficulties met in completing the forms leave their mark on the completed form; for example, it will then be clear if there is too little space to enter information.

However, before incorporating any change in a form that's in use, check out the change with all people who use the form. An improvement from one user's point of view may be a disaster from another's.

7.13. FORM CONTROL

In the organization's central control point for all forms, there should be a file for each form used in the organization. In this file should be kept a history of the form: when and by whom it was initiated, the modifications that have been made to it, and its printing history (how often it has been ordered, in what quantities, from whom, and at what price).

To control costs, all orders for replenishment of stock must be cleared through the central control point.

The central control point must have the sole authority to authorize the issue of a new form. In this way such important questions as the following can be answered.

1. Is this form necessary, or is there already in existence another form that does the same thing?
2. Can this form be combined with an already existing form?
3. Does this form comply with company standards with respect to margins, tab stops, form size, and so on?
4. Are we ordering in economical quantities?

The obstacle on which most central control of forms founders is the unwillingness of organizations to invest in the staffing and enforcement of the effort. Running the central control requires both knowledge and the pay level commensurate with this knowledge. It also requires dedicated and motivated personnel. This situation exists only if use of the central control is enforced. Attempts to control in an environment that allows a plethora of inadequately designed, overlapping, and unauthorized forms to spring up results in demoralization of a knowledgeable, highly paid staff.

EXERCISES

1. Where is the best place for instructions on how to fill out a form?
2. How can you tell whether you've allowed enough space on a form to allow the entry of the requested information?
3. Name three common failures in the specification of form content.
4. Show a layout of how you'd ask for the following information on a form.
   a. An applicant's name and address
   b. An applicant's sex
   c. The date on which an order is made
   d. The amount of a check
5. Name two methods of avoiding the need to supply the same information on forms more than once.
6. Describe one method of facilitating the routing of copies of a multipart form.
7. What is the preferred approach to selective reproduction of information on a multipart form?
8. If a form is laid out on both sides of a sheet of paper, what must the front of the form contain?
9. What are the three principal factors to be considered when determining form layout?
10. Suppose a form is to be completed on a typewriter. List the layout considerations imposed by this requirement.
11. Who is responsible for form content?

12. Suppose a form is to be bound in a binder. List the layout considerations imposed by this requirement.
13. How can you be sure you've laid out a form so the blank form can be reproduced?
14. What should you do before incorporating a suggested change into a form currently being used?
15. Give two reasons why all forms should be controlled from a central point.
16. If the same information is to appear on more than one form, how would you "interlock" the forms?
17. What is the purpose of an input form?
18. What is the prime consideration in input form layout?
19. What is the best form size? Why?
20. If instructions for filling out a form are physically separated from the form itself, how can you relate the instructions to the form?
21. Who is responsible for form layout?
22. Smith's Restaurant offers twelve different entrees for lunch. It also offers seven different appetizers, but it finds that over 75 percent of its appetizer orders are either soup du jour (the only soup offered) or tomato juice (the only juice offered). Three different beverages (coffee, tea, and milk) are available, and dessert may be ordered from a selection of ten. The entrees, desserts, and appetizers other than soup du jour and tomato juice change from day to day.

    The city in which the restaurant is located levies a sales tax on all restaurant food.

    It's the reastaurant's policy to give each customer an individual check and to have customers pay the waitress. The complete check, including the calculations, is made out by the waitress. The check isn't machined by a cash register or any other device.

    The restaurant is primarily concerned with offering its customers good food in pleasant surroundings. Poshness is not a concern. The restaurant doesn't honor credit cards.

    Lay out a check form the waitresses will use for recording the customer's lunch order. (Don't be concerned with advertising, etc.)

23. The following is a request-for-change procedure.
    a. All requests for change must be made in writing by filling out a request-for-change form. This form requires the documentation of the following information.
        i. A description of the requested change
        ii. A listing of the benefits to be derived from the change
        iii. A classification of the change into one of the following categories
            (1) Changes that must be immediately incorporated into the design
            (2) Changes that are desirable but that can wait for implementation until the initial system is installed
    b. Postponable changes are tabled until after the initial system is installed.

c. Changes that require immediate attention are handled in the following way.
   i. The project leader spells out the impact of the change on the system and the project and estimates the change in schedule and budget required to institute the change in specifications.
   ii. This information is transmitted to the user.
   iii. Implementation of the specifications change isn't begun until the user approves the changes in schedule and budget.

The completed forms are to be bound in a standard three-ring binder.
Lay out a form to be used with this procedure.

24. A company's biweekly paychecks are to be printed on a computer high-speed printer, which prints 132 characters on a line. At present no one's gross pay for one pay period exceeds a three-figure dollar amount. Federal, state, and FICA tax are withheld from everyone's pay. Most employees belong to both the company's group insurance and retirement plans, both of which are contributed to by the employees. At present there are six other possible payroll deductions; however, few employees incur any of these deductions, and if they do, they hardly ever incur more than one or two at a time. All paychecks are distributed to the 8,000 employees by their department supervisors. Departments are identified by a four-digit number. Lay out the check and earnings/deductions form.

8

Terminal Dialogue Design

The *terminal* is the instrument through which a person, the operator, interfaces with a data communications system. The operator uses the terminal to carry on a dialogue with the system. Each time the operator enters information into the system via a terminal, he or she expects the system to respond, also through the terminal.

The degree to which a system is used depends on the extent to which the operator finds the dialogue rewarding. Thus the object of terminal dialogue design is to avoid frustrating the operator.

A good dialogue design strives to reach at least the following seven goals.

1. *Communication.* The operator understands what the system is saying, and the system understands what the operator is saying.
2. *Minimum operator action.* The action required of the operator is the minimum possible. For example,
   a. If the operator is to select one of several options, the options are displayed, and the operator then selects one with a minimal operation, such as pointing to one with a light pen or entering a single character from the keyboard. It's not necessary to enter a description of the option chosen.

   b. If the operator must repetitively enter a sequence of data items, this can be done without entering any other information to indicate what the operator wants to do.
   c. If the operator is to enter a curve, it can be drawn with the light pen. It's not necessary to provide any other descriptive input.
3. *Low special-skill requirement.* The way the terminal is used is apparent to the operator. It's not necessary to acquire special skills in order to use the terminal.
4. *Standardized operation.* The operator always performs similar actions the same way. For example,
   a. An option is always selected in the same way; this might be by means of a light pen, or by making an entry from the keyboard. It's not necessary to do it one way sometimes and another way other times.
   b. If abbreviations are used, only one abbreviation is used for a given word, and that word is always abbreviated. The operator doesn't have to use an abbreviation one time and the whole word another time.
5. *Stability.* It isn't possible for the operator to get into trouble. Operator mistakes don't go undetected, and the operator is told what he or she did wrong and how to correct the mistake.
6. *Satisfactory reponse time.* The response to the operator's input is rapid enough to avoid making the operator mark time at the terminal or disrupt his train of thought.
7. *One-hand operation.* The operator may want to operate the terminal with both hands, and it should be possible to do so. But it isn't necessary to operate with both hands, because one hand may be needed to turn pages or use a telephone. Thus, for example, use of a shift key that doesn't lock is avoided.

The degree to which a dialogue can realize these goals depends on the character of the dialogue. Four factors influencing the character of a dialogue are:

1. The source of the input to be entered from the terminal
2. The need to describe conditions and actions
3. The cost of the network required to support the dialogue
4. The availability of dialogue-program generators

Examples of the influence of input source on the character of the dialogue are:

1. *Batch data entry.* The operator is entering input from a stack of forms. As a consequence,
   a. The dialogue should allow the data to be entered in the same sequence as it appears on the form.
   b. It should be possible for the operator to enter into the dialogue without looking at the display. In this way the operator can keep his or her attention on the forms while reading the data to be entered.
2. *Realtime inquiry.* The operator is using the terminal to get information to be used in an ongoing conversation. As a consequence, it should be possible to get the information rapidly; the data that must be entered to identify the information wanted should be minimal, probably a code.
3. *Realtime data entry.* The operator is using the terminal to enter data that's being furnished by a person with whom the operator is conversing. As a consequence, the operator must be able to enter the information in whatever order the person furnishes it. Therefore, the operator must precede each unit of information with an action code that identifies the information to follow.

Inquiries can also be made in a nonrealtime mode. Under such circumstances, the operator doesn't enter a code to indicate what information is wanted. Instead, the system tells the operator what inquiries are possible, and the operator selects one.

Dialogues for batch entry and nonrealtime inquiry can be highly structured and consequently do not have to require highly specialized operator skills. Realtime inquiry and data entry, with the use of action codes in a relatively unstructured situation, require the development of significant specialized operator skills.

A typical situation in which the need to describe conditions and operations arises is in a management information system. The desire is to get the system to select certain classes of data from an existing data base, manipulate this data in various ways, and then present the results of the data manipulation. To describe this desire to the system, the operator must specify conditions and operations. Consequently the dialogue must include programlike statements.

As any programmer knows, despite the development of application-oriented languages, use of a programming language requires training and practice. Moreover, programming languages are inherently unstable.

Generally speaking,

1. The fewer actions the system requires the operator to take
2. The less the skill required of the operator, and
3. The more stable the dialogue

then the more characters the system must display. And the more characters, the more costly the network. Thus dialogue design is frequently a compromise between quality dialogue and economically feasible systems.

To support a dialogue, programs must be written. General-purpose generators will be developed to generate these dialogue-supporting programs, and these generators will place restrictions on the dialogues the generated programs will support. Nevertheless, it will be desirable to use these generators and put up with the restrictions, for the same reasons that it's desirable to use COBOL and put up with its restrictions.

8.1. TERMINAL CHARACTERISTICS

Some terminals are custom designed for a particular application, such as shop-floor data collection. However, custom designed terminals have two disadvantages.

1. The first design of a system in generally not the last. If a terminal is customized on the basis of an initial design, it becomes difficult to introduce improvements when the need for them subsequently becomes apparent.
2. Although a terminal may initially be installed for just one application, normally other applications are installed later on the same terminal. Relative ease of operation for the first application is generally purchased at the price of inconvenience of operation for subsequent applications.

A common general-purpose terminal input facility is a keyboard.

A common general-purpose terminal output facility is a printer. The main disadvantage of a printer is its slowness. It's also basically limited to alphanumeric output. Some people also complain about its noise.

Another common general-purpose output facility is a cathode ray tube used as a display device.

If the output facility is a display device, a light pen can be used as an input facility.

This chapter assumes a general-purpose terminal with a keyboard and a light pen as input facilities and a display device as an output facility. It also assumes that if printed output is required, it can be provided as an optional feature of the terminal. However, most of the principles stated here can be applied, with or without modification, to a terminal with almost any other combination of characteristics.

8.2. OPERATOR TYPES

There are two extreme operator types: the *dedicated operator* and the *casual operator.* Most operators fall somewhere on a continuum between these two extremes.

The more time the operator spends at the terminal, the more dedicated he or she is. The operator who works all day at the terminal is a fully dedicated operator. A data entry operator is an example of a fully dedicated operator.

The operator may not be fulltime, but use of the terminal may be instrumental to his job. An airlines reservations clerk is an example of such an operator. This factor also tends to increase the dedication of the operator.

The opposite of the dedicated operator is the casual operator. The less frequently the operator uses the terminal and the less instrumental terminal use is to the operator's job, the more casual he or she is.

The dedicated operator gets relatively frequent practice using the terminal in a situation in which immediate feedback is supplied on the operator's performance. Such an operator can be expected to become skillful in the use of the terminal and also tends to be willing to undergo a lengthy training program in how to use the terminal.

The typical casual operator attains at best a limited ability to use the terminal and usually balks at the idea of a training program that consumes any significant amount of time. Consequently the actions required of the casual operator must be minimal and highly standardized, no special skills can be demanded, and the dialogue such an operator uses must be very stable.

The actions of the dedicated operator should also be minimal and standardized, but if the character of the dialogue requires it, the dedicated operator can be expected to acquire a high level of specialized skill in both using the dialogue and in recovering from errors. Thus the use of codes and programlike statements becomes possible.

8.3. USE OF NATURAL-LANGUAGE DIALOGUE

One of the goals of good dialogue design is to keep the requirement for special operator skills as low as possible. A natural reaction to the articulation of this goal is to say, "OK. Let's model the dialogue after the natural language. Everybody knows how to use that."

Unfortunately, natural-language dialogue runs afoul of another goal of good dialogue design—communication. Computers simply can't be programmed to understand all the forms of expression a natural language allows.

The next possibility that might be considered is to use a pseudonatural language such as COBOL. Here the dialogue is actually highly structured, even though it resembles natural language.

The disadvantage of a pseudonatural-language is that it's verbose. Spoken input to a system is still not technologically feasible, so all those long words, phrases, and sentences must be keyboarded. This violates the goal of minimizing the amount of operator action.

The reaction to this objection may be to say "OK. Let's make the system sensitive to only the key words in the dialogue and allow the use of abbreviations for these key words. Then the casual operator, who has little special skill but relatively more time, can use the language with no abbreviations and all the noise words, while the dedicated operator, with special skill but relatively little time, can use only the abbreviations of the key words." The difficulty with this approach is that sooner or later the casual operator will find out that shortcuts are possible and will want to use the shortcuts too. But the casual operator, not knowing what the key words and acceptable abbreviations are, may be unable to get the system to respond in the desired way. In other words, the dialogue becomes unstable.

The conclusion? Don't use anything that even smacks of natural-language dialogue. The dialogue must be completely structured. You make the inflexibility of the dialogue palatable to the operator by:

1. Minimizing the amount of operator action required
2. Standardizing the action so the operator becomes familiar with it rapidly

8.4. DIALOGUE REQUIREMENTS AND TYPES

All dialogues must provide two operator facilities.

1. The operator must be able to make decisions. He must be able to tell the system what he wants it to do.
2. The operator must be able to convey information to the system. The system may store the information, use it to carry out operations requested by the operator, or both.

One way to classify dialogues is on the basis of whether the system leads the operator through the dialogue or whether the operator leads the system through the dialogue. The first type of dialogue is *system structured,* the second *operator structured.*

A system-structured dialogue provides the operator with the facility to make decisions by displaying the options that are open and letting the operator select one. This is called *menu selection.* If the number of options is large, they can be arranged in a hierarchy, and the operator can make a choice by selecting a particular branch of the tree. At the least, the options can be grouped alphabetically.

If the information the operator has to convey is limited, menu selection may also be used to provide this facility. This is analogous to a multiple choice question.

The other basic way in which a system-structured dialogue provides the operator with the facility to convey information is a technique called *form filling.* This is analogous to fill-in-the-blank questions. The technique is identical to the use of a paper form. The "form" is displayed on the screen with both the captions and the blanks shown. The blanks are shown as a series of underscores. The operator enters data from the keyboard, and as each character is entered, it replaces the appropriate underscore on the screen.

System-structured dialogue is a must for the casual operator. It's often appropriate for the dedicated operator as well, but there are instances in which it is convenient for the dedicated operator to do the structuring of the dialogue.

In a system-structured dialogue, the options available to the operator are displayed as a menu, and the operator selects one, perhaps by keying in a single character associated on the display with the option to be selected. This character can be called an *action code.*

In an operator-structured dialogue, the assumption is that the operator knows the options and their associated action codes. Consequently an option is selected simply by keying in the appropriate action code.

Form filling works fine for supplying information, as long as information is provided to the operator in the same sequence as the system asks for it. However, if this sequence is determined by some external factor, it becomes necessary for the operator to determine the sequence in which the information is entered. This can be done by entering both the caption and the information. To reduce the amount of information the operator has to key, when the information-conveying dialogue is structured, the captions are reduced to action codes.

To keep action codes short, a hierarchy of codes can be created. At the top of the hierarchy, each action code identifies the application to which the subsequently used action codes apply. Then each application can assign its own meaning to a common set of action codes.

As a memory aid, a template can be placed on the keyboard to associate action codes with keys. Each application can have its own template, and the templates can be interchanged with the switch from one application to another.

To participate in an operator-structured dialogue, the operator must be trained on how to initiate the dialogue. Thus it's appropriate only when the operator is dedicated, willing to undergo a training program, and in a position to maintain skills through frequent use.

8.5. CHARACTERISTIC OPERATIONS

All dialogues are facilitated through the incorporation of certain characteristic operations. Many terminals incorporate keys to implement these operations. If this isn't the case on the terminal you're working with, you should implement these operations.

START

An operator who wants to use a terminal must notify the system of this.

The START operation should be simple, ideally a single action such as the depression of a key. In any case, the actions required to perform the START operation should be clearly posted near the terminal so no operator can have any difficulty performing the START operation.

HELP

Once having started the terminal, the operator should be able to get the system to explain how to use the dialogue. By a simple operation, such as depressing a HELP key, the operator should be able to initiate a training program that will help with any difficulties the operator is having with the dialogue.

The operator should never have any difficulty performing the HELP operation. The best approach is to post the actions required to perform the HELP operation in the same place where the actions required to perform the START operation are posted.

LOCATION RETURN

The operator should be able to initiate the HELP operation at any point in the dialogue. After learning what he or she wants to know, the operator should be able to perform the LOCATION RETURN operation and return to the point the dialogue was at before the HELP operation was initiated.

FIELD FORMATING

As the operator enters data at the terminal, the information should automatically be formated on the screen.

RELEASE

If the terminal is buffered, the operator needs a way to have the information in the buffer released to the system. This is done by performing the RELEASE operation.

CURSOR MOVEMENT

If the terminal is buffered, the operator will want a way to correct information in the buffer before performing the RELEASE operation. This is typically done by backspacing the cursor to the character position in which the correction is to be keyed. Some cursor movement operations that are often needed are:

BACKSPACE ONE CHARACTER AT A TIME

FORWARD SPACE ONE CHARACTER AT A TIME

BACKSPACE ONE FIELD AT A TIME

FORWARD SPACE ONE FIELD AT A TIME

When moving the cursor one field at a time, it's customary to locate the cursor in the first character position of the field.

CANCEL

If the terminal is buffered and the operator has some corrections to make, it's sometimes simplest to abandon the information currently in the buffer and just start over again. Rather than use multiple cursor movement operations to position the cursor back to the first character position of the first field, it's convenient for the operator to be able to achieve this goal with a simple CANCEL operation.

FRAME MOVEMENT

The operator may want to move several frames back in the dialogue, either to review the dialogue or to change it. Some frame movement operations that are often needed are:

BACKSPACE ONE FRAME AT A TIME

FORWARD SPACE ONE FRAME AT A TIME

DELETE

After having backspaced to a given frame, the operator may decide to delete the frame. The DELETE operation allows him to do so.

INSERT

The operator may want to insert a frame between two previously entered frames. The customary operation is to backspace until the first of the two frames is displayed and then perform the INSERT operation. This operation tells the system that the frame the operator is now going to enter is to be inserted following the frame being displayed.

INTERRUPT

The operator may want to suspend a dialogue temporarily. This is done by performing the INTERRUPT operation. After the dialogue has been INTERRUPTed, another dialogue can be started at the same terminal.

RESUME

At some time subsequent to INTERRUPTing a dialogue, the operator will want to resume the dialogue where it left off. The RESUME operation tells the system to do this.

TERMINATE

At some point the operator will want to terminate the dialogue. This is done by performing the TERMINATE operation.

8.6. GENERAL DESIGN PRINCIPLES

The following are some general principles that will enhance the acceptability of the dialogue you design.

1. *One idea per frame.* All the information on one frame should pertain to one idea. When the idea changes, change the frame.
2. *Small amount of information at a time.* Display only a small amount of information on the screen at a time.

In particular, if form filling is being used, instead of displaying all captions at once, display only one caption. When the operator has entered the information called for by that caption, then display the next caption. And so on.

3. *Summarize for the operator.* The operator enters information one item at a time. When the operator finishes entering all the items of information related to a given subject, the system should present the operator with a summary of the information.

For example, if in a form filling dialogue the captions are being displayed one at a time, then when a new caption is displayed, the old caption and information remain on the screen.

As a result, captions and related information are built up on the screen until, after the operator has entered all the information related to a given subject, the screen shows all the captions and entered information on the subject. The operator can then RELEASE the screen, and the process begins again.

It's even more important to perform this summary function when the operator is using action codes to enter information in an operator-determined sequence. Here all the information related to a given subject may be entered in any sequence. The system should display the information as it's entered, so the operator can see what he's doing, but when it's time to summarize, the system should "clean up" the screen by presenting the summary in a standard format.

4. *Keep the operator informed on his actions.* Summarizing is one way of keeping the operator informed of his actions. This should also be done when menu selection is being used.

If the operator is to select several items from a menu before the option he's choosing is defined, check off each item on the menu as he selects it.

After the operator has selected an option from a menu, the frame typically changes to allow the next step in the dialogue to take place. As well as performing this function, this next frame should also display a description of the option the operator has selected.

Similarly, if the operator is using action codes, the system response should include spelling out the action called for by the code.

5. *Use clear frame-formats.*
   a. Display information in columns. Left-justify alphabetics; right-justify numerics.
   b. Display narrative in short line lengths. If necessary, display in columns instead of running the lines all the way across the screen.
   c. If appropriate, use lines of asterisks to zone the screen.
   d. Ways to draw attention to parts of a screen are:
      i. Increase the intensity.
      ii. Flash the part on and off.
      iii. Use color.

6. *Make messages from the system clear.*
   a. Use simple, grammatical, properly punctuated language.
   b. Don't use jargon.
   c. Don't use abbreviations.
   d. Don's use symbols, such as >, <, =.
   e. Don't use contractions. Also, for example, use NOT VALID rather than INVALID.
   f. Be complete. If you want the date shipped, don't ask for DATE—ask for DATE SHIPPED.
7. *Distinguish instructions from operator input.* Instructions are often displayed in conjunction with operator input. Captions in a form filling dialogue are a good example of instructions, although more extensive instructions are sometimes appropriate. Methods of making the distinction between instructions and input clear are:
   a. Separate captions from input by at least one space.
   b. Put the captions in one column, the input in another.
   c. Put the captions in parentheses.
   d. Surround more extensive instructions with asterisks.
8. *Reproduce form layout.* If the dialogue is batch data entry from forms, reproduce the form on the screen.
   a. Reproduce the form's field sequence on the screen.
   b. Put the captions in the same location on the screen as they are on the form.
      i. If they're above the fields on the form, put them above the fields on the screen.
      ii. If they're to the left of the fields on the form, put them to the left of the fields on the screen.
   c. Don't have the operator overlay the captions with the input. It's confusing.
9. *Allow the operator to anticipate the system.* A dedicated operator can often anticipate frames faster than the terminal can display them. The operator should be allowed to enter information as fast as he wants to, even ahead of the frames being displayed.
10. *Anticipate the operator.* If the information to be entered is mostly standard, display the standard information. The operator can then make any modifications required for nonstandard information and then release the frame.

If the dialogue is being used for inquiry, the system can anticipate the operator by displaying the information typically desired. Then the operator need take action only to request nonstandard information.

8.7. ERROR HANDLING

An advantage of terminal data entry is that input transcription errors can be caught as the operator makes them. The dialogue should be designed to catch as many errors as possible as they occur.

Types of error checks that can be made as data is entered are:

1. Check digit
2. Consistency
3. Reasonableness
   a. Numeric field
   b. Alphabetic field
   c. Limit
   d. Valid value (the values that can be entered for a field are limited; each value is checked to see that it's an acceptable one)
4. Mandatory field (a value for a given field must be entered; the field can't be skipped)
5. Transaction sequence (transactions must enter in a given sequence, such as receive order, start production, finish production, ship, bill, receive payment; when a transaction is entered, a check is made to see that its predecessor transactions are already in the file)

Transcription errors should be detected and the operator notified of their existence within seconds after they occur. This provides the best environment for the operator to learn what his or her errors are and how to avoid them.

Error notification should be by means of an audible signal so the operator doesn't have to search the screen continually for error warnings. Characteristically, when an error occurs, the keyboard also locks.

The audible error signal should be accompanied by a display on the screen that tells the operator what has been done wrong and what should be done to correct the error.

The drawback to immediate error notification is that the operator may grow lax, knowing that the system will catch errors anyhow. To prevent this, the system should keep a log of the operator's errors and periodically report them to the operator's supervisor. The operator should be aware that this monitoring is going on.

8.8. SECURITY

After the START operation, the system should require the operator to prove he has a right to use the system by entering a password. To minimize the likelihood that the password will be learned by unauthorized personnel, the password, unlike other input, shouldn't be displayed when it's entered and should be changed periodically.

8.9. RESPONSE TIME

One goal of good dialogue design is a satisfactory response time.

The broadest kind of response time has to do with availability of the system itself. It's possible that on some occasions when the operator wants to use the system, the system may be unavailable, either because the system is completely tied up serving other operators or because all the terminals available are in use. Some unavailability may be tolerable, but too much will discourage the operator from even trying to use the system.

The next level of response is that the system should always give the operator a response for every input the operator enters. Lack of such a response leaves the operator in doubt as to whether the system correctly received the input.

Given that the system is available a reasonable amount of the time and that the system responds to each operator action, the third level of concern with respect to response time is: How quickly must the system respond to give an acceptable response time? The only general answer to this question is that it depends on the operator action. For some actions a quick response is required; for others a slower response is acceptable.

The time it takes a system to respond to an operator action varies with the extent to which the system is being used when the action occurs. Of course, this system utilization varies. As a consequence, to talk about response time meaningfully we must talk about average response time and the standard deviation of response time. Let's talk about the standard deviation first.

Given a certain type of operator action, for which it has been determined that a particular average response time is acceptable, the actual response time shouldn't vary greatly from that average. In general, an operator can adjust more readily to an average response time that would generally be considered too slow than to wide ranges of response time. If response time does vary widely, the typical operator reaction to an unusually long response time is to pummel and bang the terminal in the belief that it's not working. To avoid such situations, the standard deviation of response time should normally be no more than half the average response time.

Some common types of operator action and the kind of average response time required are as follows.

1. *START operation.* Within one second the system should respond either with an indication that it's available to serve the operator or with a busy signal (the system has received the START signal but at the moment isn't able to serve the operator).
2. *Inquiry.* The adequacy of the speed with which the system responds to an operator's inquiry varies with the operator's circumstances. For example, an operator who's talking to a customer will want a fast response time (two seconds or less). On the other hand, a storekeeper who's sitting at a desk and making an inquiry concerning stock level may be willing to wait longer for a response, perhaps ten seconds or more. Another factor bearing on the adequacy of the response time to an inquiry has to do with the extent to which the operator has the ability to do something else while waiting for the system's response. If the operator is essentially captive at the terminal and can do nothing but wait for the system's response, response time must be relatively short: certainly no more than fifteen seconds, and preferably well below this maximum. On the other hand, if the terminal and the dialogue fit into the operator's work environment in such a way that making an inquiry doesn't prevent carrying on other work, then the response time can be quite long (more than fifteen seconds), and the operator will be able to adjust to the situation. In such a case the best approach would be to have the system give the operator a rapid indication (in two seconds or less) that the inquiry has been received and then return the actual response to the operator as soon as is practical. For example, if the dialogue is programlike, the operator will enter the program and, after receiving a rapid indication that the program has been accepted, will be willing to wait an extensive period of time, perhaps even hours, for the results of the program execution.
3. *Conversation.* If the dialogue requires the operator and the system to make a large number of message exchanges to complete a transaction, the operator must concentrate on the conversation. To allow this the system must respond to each operator message within two seconds.

4. *Batch data entry.* The system must be designed so the operator can enter data as rapidly as he desires.
5. *Scanning.* In scanning through a large number of frames to find a particular piece of information, the operator should be able to move from one screen to another in one second or less.
6. *Mechanical operation.* Responses to mechanical operations, such as the click of a key when it's depressed or the appearance of a line that's being drawn on the screen with a lightpen, must be almost instantaneous (within a tenth of a second).

8.10. NETWORK COST

One of the factors influencing the character of a dialogue is the cost of the network required to support it. In general, the more we try to attain the goals of good dialogue design, the more we increase network cost. At some point it may appear to be necessary to compromise the quality of the dialogue in order to keep the system economically feasible. The danger here is that in trying to make the system economically feasible, you may destroy its operational feasibility.

The general principle behind maintaining the quality of the dialogue without unduly increasing network cost is to distribute the intelligence throughout the network. The basic idea here is to use a terminal with processing capabilities and to divide the dialogue into two parts: the dialogue that can be handled at the terminal and the dialogue that must be transmitted to the central computer.

For example, a major part of a dialogue may be to collect a batch of data piece by piece. The entry of each piece of data involves an interchange between the operator and the system, and during each interchange the system presents a large number of characters to the user. Each time the operator enters a piece of data, it's subjected to extensive validity checking. The data collection dialogue and validity checking can all be handled at the terminal. The central computer doesn't become involved until the collected batch of data is transmitted to it.

As another example, the central computer may have to make many responses to the operator, each response involving a large number of characters. The central computer can communicate its responses to the terminal in a highly compacted, coded form, and the terminal can take over the responsibility of transmitting the responses in a form intelligible to the operator.

In this short discussion of intelligence distribution, we've considered a distribution at only two levels, the terminal and the central computer. Of course, it's possible, and sometimes feasible, to distribute the intelligence at many levels of concentration throughout the network.

8.11. DEVELOPING THE DIALOGUE

Developing the dialogue is an exercise in functional specification and design. In this section we'll concentrate on the functional specification aspects of the process.

The user of a dialogue is the operator. The first step in functional specification is to determine what the user's (operator's) problem is. The next step is to determine what will constitute a satisfactory solution to this problem. These steps determine the nature of the dialogue to be developed.

The next step is to define the dialogue: What are the input messages the operator will enter at the terminal, and what will be the system responses to these messages? This definition must be complete. It must include all operator actions and must show, in rough detail, the screen layout for the entire dialogue.

The next step is to simulate the dialogue, first manually and then with the use of programs. At each step in the simulation, difficulties with the dialogue will be uncovered. The dialogue definition must be revised to eliminate the difficulties, and the simulation steps must then be repeated.

The first manual simulation step can consist of nothing more than an operator writing a message on a card and passing it to the analyst, who plays the role of the system, writes the system's response on another card, and passes it back to the operator. At first, the "operator" should be simulated by another member of the data processing department. After the difficulties detected at this level have been removed from the dialogue, live operators, who will actually be using the system, should play the operator role.

The second manual simulation step can be to have the operator enter messages at a terminal. The messages go to the analyst at another terminal, who continues to play the role of the system by entering the system's responses into his terminal for transmission to the operator's terminal.

Finally, the dialogue can be simulated using programs. Special programs can be written to perform the simulation, or in some instances, transaction simulator programs are available.

Only after a dialogue has passed all its simulation steps should it be considered specified. The simulation should be carried out with enough different live operators to maximize the probability that all ambiguities have been eliminated from the dialogue.

8.12. THE INFORMATION CENTER

Up to now we've assumed that the person who wants to use the system will operate the terminal, and we've ignored the fact that such an arrangement may not be possible. A management information system is a good example of this situation. The users of such a system are management, the epitome of a casual operator. Yet effective use of the system requires a programlike dialogue, which calls for a dedicated operator.

One solution to this problem is an information center. When a manager wants some information, he calls the information center and describes his needs to an information specialist. The specialist operates a terminal to obtain the information and then transmits it to the manager. The manager may even have a terminal in his office, but it's used only to transmit to him information obtained by the information specialist.

An information center has several advantages. One is that use of it to get information isn't limited to the information that can be obtained from a computer system via a terminal. The manager always requests information in the same way: through the information specialist. The information specialist then generates the information in the appropriate way—by inquiring from a terminal, or by referring to computer printouts, microfilm, books, periodicals, letters, or other people.

Another advantage of an information center is economic. It may significantly reduce network cost.

EXERCISES

1. What is the object of terminal dialogue design?
2. List seven goals of good dialogue design.
3. Which are better—custom designed terminals or standard terminals? Why?
4. What two things can be expected of a dedicated operator that can't be expected of a casual operator?
5. Distinguish between system-structured and operator-structured dialogues. Describe how each handles:
   a. Decisions to be made by the operator
   b. Data entry
6. List thirteen operations characteristic of dialogues.
7. How should the operator be notified of an error? Why?
8. What is the maximum acceptable variation in system response time to an operator input?

9. What is the maximum average response time in the following situations?
   a. An operator performs the START operation.
   b. An operator talking to a customer makes an inquiry.
   c. The operator is conversing with the system.
10. What is the general approach to maintaining quality dialogue without inordinately increasing network cost?
11. When can a dialogue be considered specified?
12. Two characteristics of a management information system are that the users are managers and that the dialogue is programlike. What problem does this pose, and how is it solved?
13. In Figure 8.1 is the form of Rex-Sell Corporation's invoices. Batches of these invoices are to be keyed by data-entry clerks. The display on their terminals allows for 12 lines of 40 characters each.
   a. Design the frames to be used in this operation.
   b. Specify the sequence in which these frames are to be displayed.
   c. Specify the validity checks to be made on each input field.
   d. In addition to the thirteen characteristic operations you listed in answering Exercise 6, what other general operations should be convenient for the operator in this application?

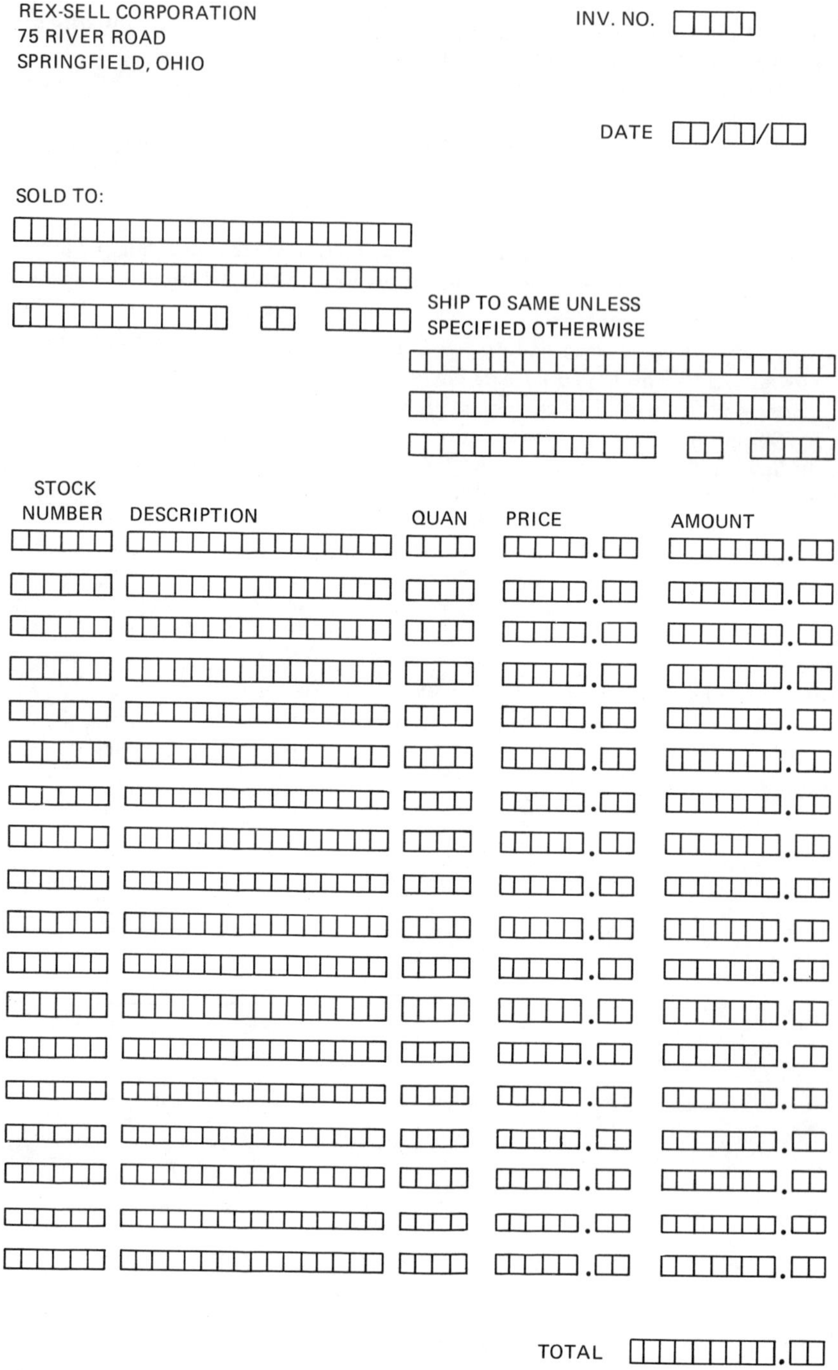

REX-SELL CORPORATION
75 RIVER ROAD
SPRINGFIELD, OHIO

INV. NO. □□□□□

DATE □□/□□/□□

SOLD TO:

SHIP TO SAME UNLESS SPECIFIED OTHERWISE

STOCK NUMBER	DESCRIPTION	QUAN	PRICE	AMOUNT

TOTAL

Figure 8.1 Invoice.

9

Decision Tables

A *procedure* specifies what actions are to be taken given various possible conditions. The following is an example of a procedure.

If the quantity ordered for a particular item equals or exceeds the minimum discount quantity *and* the order is from a wholesaler, give the customer a discount and make the shipment. This presumes that there is sufficient quantity on hand to fill the order.
If the quantity order is less than the discount quantity, bill at regular rates and make the shipment even if the customer is a wholesaler. Do the same if the sale is retail.
If there is not sufficient quantity on hand, bill as above, ship what can be shipped, and backorder the remainder of the order. It must be emphasized that, in this situation as well, even if the discount quantity is ordered, the discount is not given if the customer is a retailer.

Decision tables are a standardized way of describing procedures. They have the following advantages over alternative ways of describing procedures.

1. They require the use of standardized language.
2. They provide methodical procedures for:
   a. Maximizing the completeness of the description
   b. Eliminating all duplication
   c. Eliminating all redundancy
   d. Identifying all contradictions

The first step in developing a decision table is to identify the conditions and actions involved in the procedure. If the procedure is presently described in a narrative form, the conditions and actions can be isolated by underlining them. The conditions in the description presented above are underlined in Figure 9.1, and the actions in the description are underlined in Figure 9.2. The underlined conditions and actions can then be extracted from the description and listed, as shown in Figure 9.3.

If the quantity ordered for a particular item equals or exceeds the minimum discount quantity and the order is from a wholesaler, give the customer a discount and make the shipment. This presumes that there is sufficient quantity on hand to fill the order.

If the quantity ordered is less than the discount quantity, bill at regular rates and make the shipment even if the customer is a wholesaler. Do the same if the sale is retail.

If there is not sufficient quantity on hand, bill as above, ship what can be shipped, and backorder the remainder of the order. It must be emphasized that, in this situation as well, even if the discount quantity is ordered, the discount is not given if the customer is a retailer.

Figure 9.1 Conditions.

If the quantity ordered for a particular item equals or exceeds the minimum discount quantity and the order is from a wholesaler, give the customer a discount and make the shipment. This presumes that there is sufficient quantity on hand to fill the order.

If the quantity ordered is less than the discount quantity, bill at regular rates and make the shipment even if the customer is a wholesaler. Do the same if the sale is retail.

If there is not sufficient quantity on hand, bill as above, ship what can be shipped, and backorder the remainder of the order. It must be emphasized that, in this situation as well, even if the discount quantity is ordered, the discount is not given if the customer is a retailer.

Figure 9.2 Actions.

Conditions

1. The quantity ordered for a particular item equals or exceeds the minimum discount quantity
2. The order is from a wholesaler
3. There is sufficient quantity on hand to fill the order
4. The quantity ordered is less than the discount quantity
5. The customer is a wholesaler
6. The sale is retail
7. There is not sufficient quantity on hand
8. This situation (there is not sufficient quantity on hand)
9. The discount quantity is ordered
10. The customer is a retailer

Actions

1. Give the customer a discount
2. Make the shipment
3. Bill at regular rates
4. Make the shipment
5. Do the same (that is,
   a. Bill at regular rates
   b. Make the shipment)
6. Bill as above (that is,
   a. Give the customer a discount
   b. Bill at regular rates)
7. Ship what can be shipped
8. Backorder the remainder of the order
9. The discount is not given

Figure 9.3 List of conditions and actions.

The second step in developing a decision table is to standardize the language used in specifying the conditions and actions. Figure 9.4 shows one way of standardizing the language used in the conditions and actions listed in Figure 9.3.

Standardizing the language makes it clear that several of the conditions and actions listed are duplicates. The third step in developing a decision table is to eliminate these obvious duplications, and this is done in Figure 9.5.

It's now time to make a slightly more subtle point, still with respect to the elimination of duplication. For each of the conditions in the conditions list, it's possible to say either "Yes" or "No." Thus for the condition "Wholesale" we can say, "Yes, this is a wholesale order" or "No, this isn't a wholesale order." Similarly, for the condition "Retail" we can say, "Yes, this is a retail order" or "No, this isn't a retail order." But to say, "Yes, this is a wholesale order" is

Conditions

1. Quantity-ordered $\geqslant$ Discount-quantity
2. Wholesale
3. Quantity-ordered $\leqslant$ Quantity-on-hand
4. Quantity-ordered $<$ Discount-quantity
5. Wholesale
6. Retail
7. Quantity-ordered $>$ Quantity-on-hand
8. Quantity-ordered $\geqslant$ Quantity-on-hand
9. Quantity-ordered $\geqslant$ Discount-quantity
10. Retail

Actions

1. Bill at discount-rate
2. Ship quantity-ordered
3. Bill at regular-rate
4. Ship quantity-ordered
5. a. Bill at regular-rate
   b. Ship quantity-ordered
6. a. Bill at discount-rate
   b. Bill at regular-rate
7. Ship quantity-on-hand
8. Backorder quantity-ordered less quantity-on-hand
9. Bill at regular-rate

Figure 9.4 Standardized language.

Conditions

1. Quantity-ordered $\geqslant$ Discount-quantity
2. Wholesale
3. Quantity-ordered $\leqslant$ Quantity-on-hand
4. Quantity-ordered $<$ Discount-quantity

6. Retail
7. Quantity-ordered $>$ Quantity-on-hand

Actions

1. Bill at discount-rate
2. Ship quantity-ordered
3. Bill at regular-rate

7. Ship quantity-on-hand
8. Backorder quantity-ordered less quantity-on-hand

Figure 9.5 Elimination of duplicates.

equivalent to saying, "No, this isn't a retail order." Similarly, to say, "No, this isn't a wholesale order" is equivalent to saying, "Yes, this is a retail order." Consequently the condition "Retail" is a duplicate of the condition "Wholesale," and one of them can be eliminated. In

Conditions

1. Quantity-ordered ⩾ Discount-quantity
2. Wholesale
3. Quantity-ordered ⩽ Quantity-on-hand

Actions

1. Bill at discount-rate
2. Ship quantity-ordered
3. Bill at regular-rate
7. Ship quantity-on-hand
8. Backorder quantity-ordered less quantity-on-hand

Figure 9.6 Elimination of negative conditions.

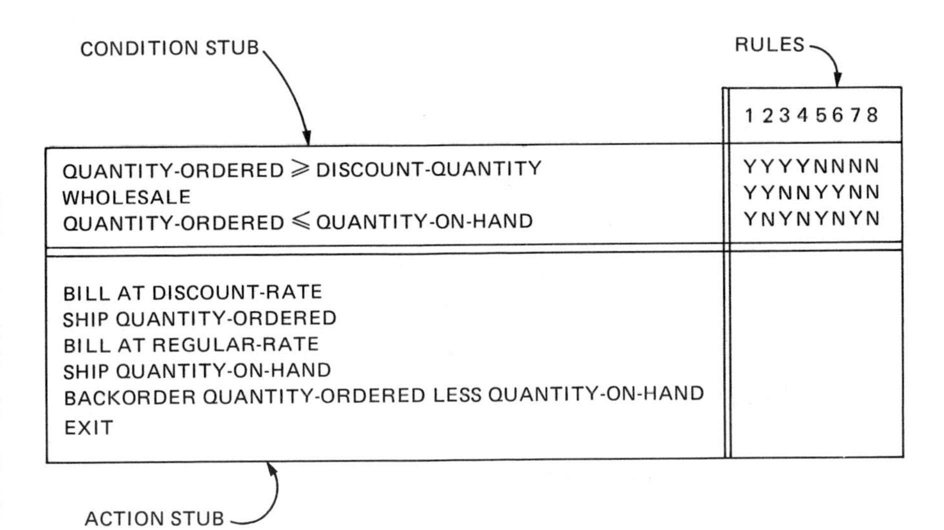

CONDITION STUB
RULES
1 2 3 4 5 6 7 8
QUANTITY-ORDERED ⩾ DISCOUNT-QUANTITY
WHOLESALE
QUANTITY-ORDERED ⩽ QUANTITY-ON-HAND
Y Y Y Y N N N N
Y Y N N Y Y N N
Y N Y N Y N Y N
BILL AT DISCOUNT-RATE
SHIP QUANTITY-ORDERED
BILL AT REGULAR-RATE
SHIP QUANTITY-ON-HAND
BACKORDER QUANTITY-ORDERED LESS QUANTITY-ON-HAND
EXIT
ACTION STUB

Figure 9.7 Decision table.

general, if one condition is the negative of another condition, duplication exists, and one of the two should be eliminated. Figure 9.6 shows our list of conditions and actions with the negative conditions eliminated.

The fourth step in developing a decision table is to put our list of conditions and actions in a decision table form, as shown in Figure 9.7.

The decision table in Figure 9.7 has one more action than does our list of actions. This is the EXIT action. Every decision table has an EXIT action, and it's always the last action in the action list.

Also shown in Figure 9.7 is some decision table terminology.

1. The conditions are listed in the *condition stub.*
2. The actions are listed in the *action stub.*
3. To the right of the condition and action stubs are a number of columns, each of which is called a *rule.*

There are as many rules in a decision table as are necessary to provide for all possible combinations of conditions. These combinations are indicated as follows.

1. If a condition must be present for a rule to hold, a Y is entered at the intersection of the condition row and the rule column. This is called a *yes entry.*
2. If a condition must be absent, an N is entered. This is a *no entry.*

Having enough rules to provide for all combinations of conditions guarantees that the procedure described in the decision table is *complete*—that is, given that all pertinent conditions are listed in the condition stub, no possible course of action is left out of the description.

It now remains to indicate the actions to be taken given the conditions in each rule. Let's look at the first paragraph in the narrative procedure description, which is reproduced below.

> If the quantity ordered for a particular item equals or exceeds the minimum discount quantity and the order is from a wholesaler, give the customer a discount and make the shipment. This presumes that there is sufficient quantity on hand to fill the order.

The conditions in this paragraph are the same as those in the first rule of the decision table. Therefore the actions described in this paragraph should be indicated in rule 1 on the decision table, as shown in Figure 9.8. Actions are indicated as follows.

1. If an action is to be taken, an X is entered.
2. If an action isn't to be taken, a dash is entered.

Figure 9.9 shows the decision table with the actions for each rule indicated.

In many procedures, actions must be taken in a specific sequence. In decision tables, this sequence is indicated by the sequence in which the actions are listed in the action stub. Our decision table doesn't have this requirement, but in Figure 9.10 we've rearranged the action order to group related actions on the theory that this makes the procedure easier to understand.

	1	2	3	4	5	6	7	8
QUANTITY-ORDERED ≥ DISCOUNT-QUANTITY	Y	Y	Y	Y	N	N	N	N
WHOLESALE	Y	Y	N	N	Y	Y	N	N
QUANTITY-ORDERED ≤ QUANTITY-ON-HAND	Y	N	Y	N	Y	N	Y	N
BILL AT DISCOUNT-RATE	X							
SHIP QUANTITY-ORDERED	X							
BILL AT REGULAR-RATE	–							
SHIP QUANTITY-ON-HAND	–							
BACKORDER QUANTITY-ORDERED LESS QUANTITY-ON-HAND	–							
EXIT	X							

Figure 9.8 Rule 1.

	1	2	3	4	5	6	7	8
QUANTITY-ORDERED ≥ DISCOUNT-QUANTITY	Y	Y	Y	Y	N	N	N	N
WHOLESALE	Y	Y	N	N	Y	Y	N	N
QUANTITY-ORDERED ≤ QUANTITY-ON-HAND	Y	N	Y	N	Y	N	Y	N
BILL AT DISCOUNT-RATE	X	X	–	–	–	–	–	–
SHIP QUANTITY-ORDERED	X	–	X	–	X	–	X	–
BILL AT REGULAR-RATE	–	–	X	X	X	X	X	X
SHIP QUANTITY-ON-HAND	–	X	–	X	–	X	–	X
BACKORDER QUANTITY-ORDERED LESS QUANTITY-ON-HAND	–	X	–	X	–	X	–	X
EXIT	X	X	X	X	X	X	X	X

Figure 9.9 Actions indicated.

	1	2	3	4	5	6	7	8
QUANTITY-ORDERED ≥ DISCOUNT-QUANTITY	Y	Y	Y	Y	N	N	N	N
WHOLESALE	Y	Y	N	N	Y	Y	N	N
QUANTITY-ORDERED ≤ QUANTITY-ON-HAND	Y	N	Y	N	Y	N	Y	N
BILL AT REGULAR-RATE	–	–	X	X	X	X	X	X
BILL AT DISCOUNT-RATE	X	X	–	–	–	–	–	–
SHIP QUANTITY-ORDERED	X	–	X	–	X	–	X	–
SHIP QUANTITY-ON-HAND	–	X	–	X	–	X	–	X
BACKORDER QUANTITY-ORDERED LESS QUANTITY-ON-HAND	–	X	–	X	–	X	–	X
EXIT	X	X	X	X	X	X	X	X

Figure 9.10 Actions rearranged.

The fifth step in developing a decision is to eliminate redundancy. Before we show you how to do this, let's discuss what redundancy is.

Look at the decision table in Figure 9.11. This table contains redundancy, because it tells you to do what the policeman says regardless of whether the light is red or green. Since your action doesn't depend on the condition of the light, information about its condition is redundant.

One way to eliminate this redundancy is to eliminate the condition "Light is red" from the condition stub. But in a table with more conditions, this condition might be pertinent for some other rules. Consequently we need another way of eliminating redundancy, and this way is shown in Figure 9.12. Here we indicate that the condition of the light is irrelevant by entering a dash (–) in the rule entries of the condition. This is called an *indifference entry.*

Redundancy is eliminated from a decision table in a completely methodical way. The first step is to detect the redundancy. Redundancy exists when all of the following hold.

1. Two rules have the same actions.
2. These two rules have the same condition entries for all conditions but one.
3. For this remaining condition, one rule has a yes entry, the other a no entry.

We can see that rules 1 and 2 in Figure 9.11 meet these criteria, and

	1234
POLICEMAN MOTIONS YOU ON	YYNN
LIGHT IS RED	YNYN
PROCEED	XX––
STOP	––XX
EXIT	XXXX

Figure 9.11 Redundancy.

	12
POLICEMAN MOTIONS YOU ON	YN
LIGHT IS RED	––
PROCEED	X–
STOP	–X
EXIT	XX

Figure 9.12 Eliminating redundancy.

so do rules 3 and 4. This proves the existence of redundancy in this decision table in a methodical way. It also indicates that the redundancy is related to the condition where the rule entries differ. In the case of Figure 9.11 this is the "Light is red" condition.

After the redundancy is detected, the next step is to eliminate it. This is done by collapsing the two rules constituting the redundancy into one rule by replacing with an indifference entry the yes entry and the no entry for the condition to which the redundancy is related. Thus rules 1 and 2 in Figure 9.11 are collapsed into rule 1 in Figure 9.12. Similarly, rules 3 and 4 in Figure 9.11 are collapsed into rule 2 in Figure 9.12.

The decision table in Figure 9.10 contains redundancy. Why don't you apply the procedure just given to identify and eliminate this redundancy? When you're finished, check the resulting decision table with ours, which is shown in Figure 9.13.

To obtain the table in Figure 9.13 we eliminated the redundancy by collapsing rules 7 and 8 into rules 5 and 6. You may have taken a different and equally valid approach to eliminating the redundancy: collapsing rules 7 and 8 into rules 3 and 4.

The general effect of eliminating redundancy is to reduce the number of rules in a decision table. Another way to reduce the number of rules is through use of the *else rule.* The else rule is just a residual category, and it is used as follows.

First, specify the rules with condition entries to cover all sets of actions but one. This final set of actions is the course of action to be taken if none of the explicitly specified conditions holds. This residual course of action is specified in the final column of the decision table, and instead of specific condition entries, this rule has the word "else" entered in the condition part of the rule. Figure 9.14 shows our decision table with the use of an else rule.

	1 2 3 4 5 6
QUANTITY-ORDERED ≥ DISCOUNT-QUANTITY	Y Y Y Y N N
WHOLESALE	Y Y N N – –
QUANTITY-ORDERED ≤ QUANTITY-ON-HAND	Y N Y N Y N
BILL AT REGULAR-RATE	– – X X X X
BILL AT DISCOUNT-RATE	X X – – – –
SHIP QUANTITY-ORDERED	X – X – X –
SHIP QUANTITY-ON-HAND	– X – X – X
BACKORDER QUANTITY-ORDERED LESS QUANTITY-ON-HAND	– X – X – X
EXIT	X X X X X X

Figure 9.13 Redundancy eliminated.

	1 2 3 4	ELSE
QUANTITY-ORDERED ⩾ DISCOUNT-QUANTITY	Y Y Y N	
WHOLESALE	Y Y N –	
QUANTITY-ORDERED ⩽ QUANTITY-ON-HAND	Y N Y Y	
BILL AT REGULAR-RATE	– – X X	X
BILL AT DISCOUNT-RATE	X X – –	–
SHIP QUANTITY-ORDERED	X – X X	–
SHIP QUANTITY-ON-HAND	– X – –	X
BACKORDER QUANTITY-ORDERED LESS QUANTITY-ON-HAND	– X – –	X
EXIT	X X X X	X

Figure 9.14 The else rule.

You can always construct a complete decision table that's free of redundancy and contradiction by following the five steps described above. However, most people find this procedure tedious and prefer to use the following approach.

As before, conditions and actions are identified, the language used in specifying the conditions and actions is standardized, and duplicate conditions and actions are eliminated. The remaining conditions and actions are then used to construct the condition and action stubs of a decision table. However, no arbitrary set of rules, determined by all the possible combinations of conditions, is constructed. Instead, as many rules are constructed as the procedure appears to call for.

If we use this approach to the order-handling procedure we've been using as our principal example in this chapter, we might come up with a decision table that looks like the one in Figure 9.15. The rules in this table are related to the procedure described at the beginning of this chapter in the following way.

Rule 1. If the quantity ordered for a particular item equals or exceeds the minimum discount quantity and the order is from a wholesaler, give the customer a discount and make the shipment. This presumes that there is sufficient quantity on hand to fill the order.

Rule 2. If the quantity ordered is less than the discount quantity, bill at regular rates and make the shipment even if the customer is a wholesaler.

Rules 3 and 4. Do the same if the sale is retail.

Rules 5, 6, and 7. If there is not sufficient quantity on hand, bill as above, ship what can be shipped, and backorder the remainder of the order.

	1	2	3	4	5	6	7	8
QUANTITY-ORDERED ⩾ DISCOUNT-QUANTITY	Y	N	N	–	Y	N	–	Y
WHOLESALE	Y	Y	N	N	Y	–	N	N
QUANTITY-ORDERED ⩽ QUANTITY-ON-HAND	Y	Y	Y	Y	N	N	N	N
BILL AT DISCOUNT-RATE	X	–	–	–	X	–	–	–
SHIP QUANTITY-ORDERED	X	X	X	X	–	–	–	–
BILL AT REGULAR-RATE	–	X	X	X	–	X	X	X
SHIP QUANTITY-ON-HAND	–	–	–	–	X	X	X	X
BACKORDER QUANTITY-ORDERED LESS QUANTITY-ON-HAND	–	–	–	–	X	X	X	X
EXIT	X	X	X	X	X	X	X	X

Figure 9.15 Alternative decision table.

Rule 8. It must be emphasized that, in this situation as well, even if the discount quantity is ordered, the discount is not given if the customer is a retailer.

After having constructed this initial decision table, the next step is to eliminate redundancy according to the procedure already described. In the case of the decision table in Figure 9.15, the procedure applies to rules 2 and 3. The decision table with this redundancy removed is shown in Figure 9.16.

But since we didn't follow a methodical procedure in constructing the decision table in Figure 9.16, the rules in the table may not be *independent* of each other. A table containing *dependent* (not independent) rules contains redundancy and/or contradiction.

Before proceeding, let us state what we mean by *contradiction.* Look at the decision table in Figure 9.17. This table contains contradiction. Rule 1 tells you to do what the policeman says and to ignore

	1	2	3	4	5	6	7
QUANTITY-ORDERED ⩾ DISCOUNT-QUANTITY	Y	N	–	Y	N	–	Y
WHOLESALE	Y	–	N	Y	–	N	N
QUANTITY-ORDERED ⩽ QUANTITY-ON-HAND	Y	Y	Y	N	N	N	N
BILL AT DISCOUNT-RATE	X	–	–	X	–	–	–
SHIP QUANTITY-ORDERED	X	X	X	–	–	–	–
BILL AT REGULAR-RATE	–	X	X	–	X	X	X
SHIP QUANTITY-ON-HAND	–	–	–	X	X	X	X
BACKORDER QUANTITY-ORDERED LESS QUANTITY-ON-HAND	–	–	–	X	X	X	X
EXIT	X	X	X	X	X	X	X

Figure 9.16 Redundancy removed.

the light. But rule 2 tells you that if the light is red, you are to obey the light and ignore the policeman.

Now, back to rule independence. Each pair of rules in a decision table exhibits independence if there's at least one condition for which one rule has a yes entry and the other a no entry. Thus we can see that the rules in the decision table in Figure 9.17 aren't independent.

If two rules are dependent, then:

1. If they both have the same actions, the table contains redundancy.
2. If they have different actions, they're contradictory.

This proves in a methodical way that the rules in the decision table in Figure 9.17 are contradictory.

	1 2
POLICEMAN MOTIONS YOU ON	YY
LIGHT IS RED	–Y
PROCEED	X–
STOP	–X
EXIT	XX

Figure 9.17 Contradiction.

There's no methodical way to eliminate contradiction from a decision table. Contradiction in a decision table indicates confusion about the procedure being described, and to eliminate the contradiction, the confusion must be cleared up.

However, if rule dependence indicates redundancy, this redundancy can be eliminated in a methodical way. Consequently, let's turn our attention to these methods.

The decision table in Figure 9.16 has several dependent rule pairs.

1. Rules 2 and 3
2. Rules 5 and 6
3. Rules 6 and 7

Moreover, all these dependent rule pairs indicate redundancy. So let's see how we go about eliminating this redundancy.

First we need some more decision table terminology.

1. If all the condition entries for a rule are either a yes entry or a no entry, the rule is a *pure rule.*
2. If any condition entry for a rule is an indifference entry, the rule is a *mixed rule.*

Thus in the decision table in Figure 9.16, rules 1, 4, and 7 are pure rules; the others are mixed rules.

There are two procedures for eliminating redundancy between dependent rule pairs. The first is:

> If one of the dependent rules is pure and the other mixed, the pure rule is contained in the mixed one.

This procedure applies to rules 6 and 7 in the decision table shown in Figure 9.16. Therefore rule 7 is redundant and can be removed. What we're saying is this: If the customer is a retailer and there isn't sufficient quantity to fill the order, we're going to bill at the regular rate, ship the quantity on hand, and backorder the rest of the order, and we don't care whether the quantity ordered exceeds the discount quantity or not. The decision table with rule 7 removed is shown in Figure 9.18.

The second procedure for eliminating redundancy between dependent rule pairs is:

> If both the dependent rules are mixed, there's at least one pure rule, common to both of the mixed rules, that can be eliminated from one of the mixed rules.

This procedure applies to rules 2 and 3 in the decision table in Figure 9.18, and to rules 5 and 6 as well. Let's see how it applies to rules 2 and 3 first.

Figure 9.19 shows our decision table with rules 2 and 3 expanded into the pure rules that make them up. It's now apparent that the pure rule that rules 2 and 3 have in common is rule 2B and rule 3B, which are the same. Consequently we can eliminate either rule 2B or rule 3B. If we eliminate rule 3B, we obtain the decision table shown in Figure 9.20.

	1 2 3 4 5 6
QUANTITY-ORDERED ⩾ DISCOUNT-QUANTITY	Y N – Y N –
WHOLESALE	Y – N Y – N
QUANTITY-ORDERED ⩽ QUANTITY-ON-HAND	Y Y Y N N N
BILL AT DISCOUNT-RATE	X – – X – –
SHIP QUANTITY-ORDERED	X X X – – –
BILL AT REGULAR-RATE	– X X – X X
SHIP QUANTITY-ON-HAND	– – – X X X
BACKORDER QUANTITY-ORDERED LESS QUANTITY-ON-HAND	– – – X X X
EXIT	X X X X X X

Figure 9.18 Redundancy removed.

	1	2		3		4	5	6
		A	B	A	B			
QUANTITY-ORDERED ≥ DISCOUNT-QUANTITY	Y	N	N	Y	N	Y	N	–
WHOLESALE	Y	Y	N	N	N	Y	–	N
QUANTITY-ORDERED ≤ QUANTITY-ON-HAND	Y	Y	Y	Y	Y	N	N	N
BILL AT DISCOUNT-RATE	X	–	–	–	–	X	–	–
SHIP QUANTITY-ORDERED	X	X	X	X	X	–	–	–
BILL AT REGULAR-RATE	–	X	X	X	X	–	X	X
SHIP QUANTITY-ON-HAND	–	–	–	–	–	X	X	X
BACKORDER QUANTITY-ORDERED LESS QUANTITY-ON-HAND	–	–	–	–	–	X	X	X
EXIT	X	X	X	X	X	X	X	X

Figure 9.19 Mixed rules expanded.

	1	2		3	4	5	6
		A	B				
QUANTITY-ORDERED ≥ DISCOUNT-QUANTITY	Y	N	N	Y	Y	N	–
WHOLESALE	Y	Y	N	N	Y	–	N
QUANTITY-ORDERED ≤ QUANTITY-ON-HAND	Y	Y	Y	Y	N	N	N
BILL AT DISCOUNT-RATE	X	–	–	–	X	–	–
SHIP QUANTITY-ORDERED	X	X	X	X	–	–	–
BILL AT REGULAR-RATE	–	X	X	X	–	X	X
SHIP QUANTITY-ON-HAND	–	–	–	–	X	X	X
BACKORDER QUANTITY-ORDERED LESS QUANTITY-ON-HAND	–	–	–	–	X	X	X
EXIT	X	X	X	X	X	X	X

Figure 9.20 Common rule eliminated.

Now we can collapse rules 2A and 2B back in to the old rule 2. The result of this step is shown in Figure 9.21.

Try applying the procedure just given to eliminate the redundancy in the dependent rule pair, rules 5 and 6, in the decision table shown in Figure 9.21. When you're done, check the resulting decision table with ours, which is shown in Figure 9.22.

The redundancy in our decision table has now been eliminated, and as usual, we can now reduce the number of rules in the table by means of the else rule. This is done in Figure 9.23.

We can also group the actions in the action stub on the basis of similarity. This is done in Figure 9.24. The decision table in Figure 9.24 is equivalent to the decision table in Figure 9.14, which is the

	1	2	3	4	5	6
QUANTITY-ORDERED $\geq$ DISCOUNT-QUANTITY	Y	N	Y	Y	N	–
WHOLESALE	Y	–	N	Y	–	N
QUANTITY-ORDERED $\leq$ QUANTITY-ON-HAND	Y	Y	Y	N	N	N
BILL AT DISCOUNT-RATE	X	–	–	X	–	–
SHIP QUANTITY-ORDERED	X	X	X	–	–	–
BILL AT REGULAR-RATE	–	X	X	–	X	X
SHIP QUANTITY-ON-HAND	–	–	–	X	X	X
BACKORDER QUANTITY-ORDERED LESS QUANTITY-ON-HAND	–	–	–	X	X	X
EXIT	X	X	X	X	X	X

Figure 9.21 Rule 2 reinstated.

	1	2	3	4	5	6
QUANTITY-ORDERED $\geq$ DISCOUNT-QUANTITY	Y	N	Y	Y	N	Y
WHOLESALE	Y	–	N	Y	–	N
QUANTITY-ORDERED $\leq$ QUANTITY-ON-HAND	Y	Y	Y	N	N	N
BILL AT DISCOUNT-RATE	X	–	–	X	–	–
SHIP QUANTITY-ORDERED	X	X	X	–	–	–
BILL AT REGULAR-RATE	–	X	X	–	X	X
SHIP QUANTITY-ON-HAND	–	–	–	X	X	X
BACKORDER QUANTITY-ORDERED LESS QUANTITY-ON-HAND	–	–	–	X	X	X
EXIT	X	X	X	X	X	X

Figure 9.22 Redundancy eliminated.

	1	2	3	4	ELSE
QUANTITY-ORDERED $\geq$ DISCOUNT-QUANTITY	Y	N	Y	Y	
WHOLESALE	Y	–	N	Y	
QUANTITY-ORDERED $\leq$ QUANTITY-ON-HAND	Y	Y	Y	N	
BILL AT DISCOUNT-RATE	X	–	–	X	–
SHIP QUANTITY-ORDERED	X	X	X	–	–
BILL AT REGULAR-RATE	–	X	X	–	X
SHIP QUANTITY-ON-HAND	–	–	–	X	X
BACKORDER QUANTITY-ORDERED LESS QUANTITY-ON-HAND	–	–	–	X	X
EXIT	X	X	X	X	X

Figure 9.23 The else rule.

	1 2 3 4	ELSE
QUANTITY-ORDERED ⩾ DISCOUNT-QUANTITY	Y N Y Y	
WHOLESALE	Y – N Y	
QUANTITY-ON-HAND	Y Y Y N	
BILL AT REGULAR-RATE	– X X –	X
BILL AT DISCOUNT-RATE	X – – X	–
SHIP QUANTITY-ORDERED	X X X –	–
SHIP QUANTITY-ON-HAND	– – – X	X
BACKORDER QUANTITY-ORDERED LESS QUANTITY-ON-HAND	– – – X	X
EXIT	X X X X	X

Figure 9.24 Actions grouped.

final decision table we arrived at by using the methodical approach to decision table development. However, they aren't identical, since the rules are in different orders in the two tables.

The rule order of the decision table in Figure 9.14 is preferred, since it allows a user to find the appropriate rule in a minimum amount of time. A table arranged in this way is called a *sorted* table.

An unsorted decision table, such as the one in Figure 9.24, can be sorted by rearranging the rules and the conditions until all the yes entries are pushed as far as possible into the top left-hand corner of the box of condition entries and the indifference entries are pushed as far as possible into the lower right-hand corner of the box. Sorting the decision table in Figure 9.24 yields the decision table shown in Figure 9.25.

EXERCISE 1

Develop a decision table for the following procedure.

There are two situations in which you might buy a particular stock: when the market as a whole is appreciating and when you have knowledge of a special situation. When the market in general is going up, you should buy the stock if it is appreciating faster than the market as a whole. Otherwise, you should look for another stock. If you have knowledge of a special situation, check the stock's recent price and volume action. If price has been increasing on expanding volume and decreasing on contracting volume, buy the stock. Otherwise, your information is probably faulty, and you should hold back. In general, you should not borrow money to buy stock. However, if the market is going up as a whole, you have knowledge of a special situation, and the stock is behaving in such a fashion as to indicate that it should be bought, buy the stock on margin. Otherwise, when the market is going up, don't worry about special situations.

	1	2	3	4	ELSE
QUANTITY-ORDERED ≥ DISCOUNT-QUANTITY	Y	Y	Y	N	
QUANTITY-ORDERED ≤ QUANTITY-ON-HAND	Y	Y	N	Y	
WHOLESALE	Y	N	Y	–	
BILL AT REGULAR-RATE	–	X	–	X	X
BILL AT DISCOUNT-RATE	X	–	X	–	–
SHIP QUANTITY-ORDERED	X	X	–	X	–
SHIP QUANTITY-ON-HAND	–	–	X	–	X
BACKORDER QUANTITY-ORDERED LESS QUANTITY-ON-HAND	–	–	X	–	X
EXIT	X	X	X	X	X

Figure 9.25 Sorted table.

9.1. THE GO AGAIN ACTION

If a series of cases is to be run against a decision table, the GO AGAIN action is used. The decision table in Figure 9.26 is designed to process a series of orders and uses the GO AGAIN action to do so.

The GO AGAIN action has a unique property. You may have noticed that, once you've decided, by means of checking condition entries, which rule applies to a particular case, you have, up to now, always moved down the rule and taken every action indicated, in sequence, until the EXIT action is reached. The GO AGAIN action is the only exception to this rule. Whenever, in executing the actions of a rule, the GO AGAIN action is indicated, the remaining actions in the rule aren't inspected. Instead, you return immediately to the first condition of the decision table to determine what to do next.

	1	2	3	4	ELSE
QUANTITY-ORDERED ≥ DISCOUNT-QUANTITY	Y	Y	Y	N	
QUANTITY-ORDERED ≤ QUANTITY-ON-HAND	Y	Y	N	Y	
WHOLESALE	Y	N	Y	–	
BILL AT REGULAR-RATE	–	X	–	X	X
BILL AT DISCOUNT-RATE	X	–	X	–	–
SHIP QUANTITY-ORDERED	X	X	–	X	–
SHIP QUANTITY-ON-HAND	–	–	X	–	X
BACKORDER QUANTITY-ORDERED LESS QUANTITY-ON-HAND	–	–	X	–	X
GET NEXT ORDER	X	X	X	X	X
GO AGAIN	X	X	X	X	X
EXIT	–	–	–	–	–

Figure 9.26 The GO AGAIN action.

	1	2	3	4	5	ELSE
ANOTHER ORDER	Y	Y	Y	Y	N	
QUANTITY-ORDERED ≥ DISCOUNT-QUANTITY	Y	Y	Y	N	–	
QUANTITY-ORDERED ≤ QUANTITY-ON-HAND	Y	Y	N	Y	–	
WHOLESALE	Y	N	Y	–	–	
GET NEXT ORDER	X	X	X	X	–	X
BILL AT REGULAR-RATE	–	X	–	X	–	X
BILL AT DISCOUNT-RATE	X	–	X	–	–	–
SHIP QUANTITY-ORDERED	X	X	–	X	–	–
SHIP QUANTITY-ON-HAND	–	–	X	–	–	X
BACKORDER QUANTITY-ORDERED LESS QUANTITY-ON-HAND	–	–	X	–	–	X
GO AGAIN	X	X	X	X	–	X
EXIT	–	–	–	–	X	–

Figure 9.27 Exiting from a loop.

Because of this special property, the presence of a GO AGAIN action in a decision table sets up a *loop* in the table. In some instances, it's necessary to build a special provision into such a decision table to get out of this loop. This is the case with the decision table in Figure 9.26. There's no way to exit from this table; as it's presently constituted, once you enter it, you circle around in it endlessly. The decision table in Figure 9.27 breaks this *closed loop* and provides for eventual exit from the table.

9.2. TABLE SIZE

The purpose of decision tables is to prepare easily understood documents describing procedures. Decision tables perform this function well until the procedure being described becomes complicated enough to make the table describing it unwieldy.

The solution to this problem is to subdivide decision tables that are too large into a number of smaller decision tables. This subdivision is accomplished by assigning names to decision tables and connecting the tables through use of the PERFORM action.

Figure 9.28 shows two tables, named MAIN and ORDER-PROCESS. The name of a decision table appears in the top left-hand corner of the table, just above the condition stub.

When, in executing the actions in a decision table, a PERFORM action is indicated, you stop executing these actions and go the top of the decision table named in the PERFORM action. Thus in the table MAIN in Figure 9.28, when the PERFORM action is executed,

MAIN	1	2
ANOTHER ORDER	Y	N
GET NEXT ORDER	X	–
PERFORM ORDER-PROCESS	X	–
GO AGAIN	X	–
EXIT	–	X

ORDER-PROCESS	1	2	3	4	5	ELSE
ANOTHER LINE-ITEM	Y	Y	Y	Y	N	
QUANTITY-ORDERED ≥ DISCOUNT-QUANTITY	Y	Y	Y	N	–	
QUANTITY-ORDERED ≤ QUANTITY-ON-HAND	Y	Y	N	Y	–	
WHOLESALE	Y	N	Y	–	–	
GET NEXT LINE-ITEM	X	X	X	X	–	X
BILL AT REGULAR-RATE	–	X	–	X	–	X
BILL AT DISCOUNT-RATE	X	–	X	–	–	–
SHIP QUANTITY-ORDERED	X	X	–	X	–	–
SHIP QUANTITY-ON-HAND	–	–	X	–	–	X
BACKORDER QUANTITY-ORDERED LESS QUANTITY-ON-HAND	–	–	X	–	–	X
GO AGAIN	X	X	X	X	–	X
TOTAL THE ORDER	–	–	–	–	X	–
EXIT	–	–	–	–	X	–

Figure 9.28 Performing tables.

you go to the table ORDER-PROCESS and begin following the procedure described by the table ORDER-PROCESS.

When a table is performed, you continue following the procedure described in the table being performed until the EXIT action is reached. You then return to the action following the PERFORM action that sent you to the table being performed. Thus after you enter the table ORDER-PROCESS as a result of executing the PERFORM ORDER-PROCESS action in the table MAIN, you remain in the table ORDER-PROCESS until you execute the EXIT action in that table. You then return to the action following the PERFORM ORDER-PROCESS in the table MAIN, which is the action GO AGAIN.

If you can't fit a decision table on an 8-1/2″ by 11″ sheet of paper, subdivide the table through use of the PERFORM action.

Give your decision tables meaningful names, so reading of PERFORM actions makes sense.

A decision table doesn't have to contain conditions. It can consist of an action stub alone.

9.3. THE GO TO ACTION

The GO TO action is similar to the PERFORM action in that, when you find a GO TO action in a sequence of actions that you're executing, you stop executing those actions and go to the top of the decision table named in the GO TO action. The GO TO action differs from a PERFORM action in that there's no provision for a return to the action following the GO TO action. The GO TO action unconditionally transfers your attention to the named decision table.

There's only one situation in which you should use a GO TO action. Consider the situation in Figure 9.29, which is a perfectly standard one. However, suppose that condition B depends on the outcome of action D. Then the situation has to be described as shown in Figure 9.30.

TABLE -X	1 2 3 4
CONDITION A	Y Y N N
CONDITION B	Y N Y N
ACTION C	X X – –
ACTION D	X X X X
GO AGAIN	X – X –
EXIT	– X – X

Figure 9.29 Decision table.

TABLE-X	1 2
CONDITION A	Y N
ACTION C	X –
ACTION D	X X
PERFORM TABLE-Y	X X
EXIT	X X

TABLE-Y	1 2
CONDITION B	Y N
GO TO TABLE-X	X –
EXIT	– X

Figure 9.30 The GO TO action.

EXERCISE 2

Develop a decision table for the following procedure.

A file of balance records is in order by part number. Each balance record contains the following fields.

Part-number
Description
On-hand-quantity
On-order-quantity
Available-quantity
Reorder-point
Reorder-quantity
Usage-this-year

An addition file is run against this balance file. The addition file is also in order by part number.

A change file is also run against this balance file. The change file is also in order by part number. There are four types of change records.

Receipt
Issue
Part number change
Deletion

Change records for the same part number are in the order indicated by the list above.

If a part number of a balance record is changed, it's deleted from the balance file and written on a transfer file.

Receipt and issue changes carry a quantity field. A receipt causes an increase in the on-hand-quantity and a decrease in the on-order-quantity. An issue causes a decrease in the on-hand-quantity and available-quantity and an increase in the usage-this-year.

When all changes for a given balance record have been applied, if the balance record isn't deleted, then if the available-quantity isn't more than the reorder-point, an order record is written. The order record contains the following fields.

Part-number
Description
Quantity-ordered

The quantity-ordered is the reorder-point plus the reorder-quantity less the available-quantity. After the quantity-ordered has been calculated, it's used to increase the on-order-quantity and the available-quantity.

9.4. OBSERVATIONS

The last exercise you completed is a fairly practical data processing function. Notice that in practice there is hardly ever an opportunity to make productive use of the else rule.

Also notice that if you conform to the restriction on the use of the GO TO action that this chapter proposes, the decision tables you construct automatically organize your procedure logic in a "structured programming" format.

Finally, the requirement to keep decision tables down to a manageable size forces a solution expressed in decision table terms to reflect the hierarchy of functions making up the solution.

9.5. OTHER PROCEDURE DESCRIPTION TECHNIQUES

We believe that decision tables are the best procedure description technique available. We don't think there are any other real alternatives. Nevertheless, we've been strongly urged by prepublication reviewers to consider other techniques. Certain other techniques may be fine for the purposes for which they've been designed, but we are convinced that they're not as good a procedure description technique as decision tables. In any case, here are some of the other techniques.

9.5.1. Narrative

Narrative lends itself to incompleteness, contradiction, and ambiguity. It was these characteristics that led to development of decision tables in the first place. Narrative is fine for many communications purposes, but it's almost invariably defective as a procedure description technique.

9.5.2. Flowcharts

With the advent of structured programming, it has become clear that flowcharts are a defective instrument for organizing logic; they tend to focus concentration on process rather than structure. In addition, it has long been known that flowcharts, from their nature, encourage—that is, to all intents, force—the use of abbreviations, to the detriment of their value as a communication tool. To be blunt about it, they don't communicate. Finally, flowcharts provide no mechanical check against incompleteness, contradiction, and redundancy.

9.5.3. Hierarchy Charts

Hierarchy charts are a fine tool for emphasizing those aspects of system structure that have the most impact on effective system design. However, as we've already pointed out, decision tables do form a hierarchy, and we feel they're superior to hierarchy charts, at least as a documentation tool. Like all charting techniques, hierarchy charts are an almost irresistible force toward abbreviation. As a result, they're usually incomprehensible. Therefore we believe that, if hierarchy charts are to be used to show logic in functional specifications, then each box on the charts should be backed up by either a decision table or, as a very poor minimum, a narrative description of the processing represented by the box.

9.5.4. Warnier Diagrams

A *Warnier diagram* is a chart drawn with a technique that was developed by Jean Dominique Warnier and is described in his book *Logical Construction of Programs* (New York: Van Nostrand Reinhold, 1974). Warnier's main point apparently is that a program should be structured and that it's structure should reflect the structure of the data it processes. (This is the same point Michael Jackson emphasizes in his *Principles of Program Design* (Academic Press, 1975).) Warnier diagrams are a way to depict this structure. Warnier also shows how his diagrams can be derived from truth tables (from which, incidentally, decision tables are also derived). Yourdon has made the point that, while Warnier's and Jackson's idea is good for structuring programs after the data has been structured, it's of little use when part of the problem is how the data is to be structured, which is the case in system design. In any case we will evaluate Warnier diagrams only from the perspective of how well they document procedures.

From this point of view we think Warnier diagrams fail on two counts.

1. They're designed to describe structure, not procedure. As a consequence, they make a poor procedure description tool. When he wants to show procedure, Warnier falls back on flowcharts.
2. Like flowcharts, Warnier diagrams by their nature encourage the use of abbreviations.

9.5.5. Pseudocode

Pseudocode was originally developed to give people, who wanted to restrict their tools for logical expression to the Böhm-Jacopini structures, a coding language that would do just that. For a description of pseudocode see *Assignment 4* of the *Structured Programming Textbook* published by IBM (SR20-7149).

While Böhm and Jacopini have proved that any logical operation can be described in terms of their set of structures, the structures themselves are too minimal to allow an uncontorted description of procedure. Thus IBM suggests addition of the case structure, and Yourdon has suggested the break and cycle structures, in addition to Böhm and Jacopini's fundamental three.

In any case, regardless of the number of basic structures finally fixed on, pseudocode looks much like any other code. We don't mean it isn't structured–it is. We mean that pseudocode isn't any easier to read than any other code. It may be all right for *ACM Journal* articles, but it will repel the bulk of users.

To be graduated to the rank of a vehicle for expressing procedure logic in functional specifications, pseudocode will have to stop looking so much like code and start looking more like a language. As a matter of fact, this process is already starting. Yourdon, in his course on structured analysis, is pushing what he calls "structured English" as the vehicle for expressing procedure logic. The process of specifying "structured English" hasn't ended, but already it's apparent that the structure is going to be so loose that it will be virtually impossible to determine whether a procedure description expressed in structured English is complete and free of contradiction–unless, of course, you were to convert it to a decision table.

9.5.6. Nassi-Schneiderman Charts

Nassi-Schneiderman charts spring from the same source as psuedocode, except that here the desire was to develop flowcharting conventions that would restrict logic to the Böhm-Jacopini structures. Nassi and Schneiderman rose to the challenge only too well, and as a result, their charts are useless as a practical logic documentation tool. To our knowledge no one has suggested how to add to Nassi-Schneiderman charts Yourdon's cycle structure, which seems to be the minimum to make the Böhm-Jacopini structures a practical procedure description tool. However, if you want to read about Nassi-Schneiderman charts, you can find them described in the August 1973 *SIGPLAN Notices*.

If you must use flowcharts to document procedures, we suggest that you adopt the following conventions.

1. A module is made up of an entrance, an exit, and a number of operations. Only one arrow leaves the entrance, and only one arrow enters the exit.
2. A module has a maximum of one connector, which occurs immediately after the entrance. If there's a maximum of one module per flowchart page, this connector can be numbered with the page number.
3. An operation may be:
   a. A unitary operation, which may be:
      i. A data manipulation operation
      ii. An input-output operation
      iii. A perform operation
   b. A series of operations
   c. An if-then-else
   d. A do-while
   e. A do-until
4. Arrows must either:
   a. Merge with another arrow or
   b. Lead to:
      i. An operation
      ii. The single connector
      iii. The exit
5. Arrows may not cross.
6. There may be page connectors, but they may not carry out any function other than to connect pages; that is, if pages were large enough, the page connectors would be superfluous.

Incidentally, with the exception of conventions 2 and 4.b.ii, these conventions are identical to the Böhm–Jacopini structures. Conventions 2 and 4.b.ii, taken together, are identical to Yourdon's cycle structure. All the conventions together provide the same flexibility and restraints for documenting procedures as do decision tables. Of course, they don't give you any mechanical way to check for completeness and lack of contradiction. The solution to exercise 2, shown in decision table form in Figure A.11 (in the Answer Section), is shown in a flowchart form conforming to these conventions in Appendix A.

Part 4

OTHER SKILLS

Interpersonal Relations

All the analyst's responsibilities, including data collection and communication, require the analyst to work with people continually. Project management requires the ability to get along with people. If the analyst is the first to see that there's a better way of doing things, he or she must be personable enough to convince the user of this. When the user keeps coming up with solutions as system proposals, the analyst must exercise tact in insisting that, instead of implementing the proposed system, time be spent unearthing the underlying problem motivating the user to propose solutions.

The analyst is intimately involved in all aspects of the acceptance testing used to demonstrate the acceptability of the newly constructed system. This involvement ranges from the specification of the test (these acceptance test specifications are an essential part of the functional specifications) through the generation of the test data and predetermined results to the coordination of the user, the system development team, and the computer center in test running. The job requires the utmost in tact and persistence, since the people involved, even if they are philosophically committed to the idea of an acceptance test, may tend to put off the extensive work involved in favor of more immediate demands on their time.

Thus, we can see that the ability to work successfully with people has a large bearing on the analyst's success or failure. This subject is addressed in Chapter 10, "Interpersonal Relations," the first chapter of Part 4.

Creative Thinking

In specifying the objectives to be met by a satisfactory problem solution, in identifying the primary functions of the user's organization, and in developing the functional specifications, the analyst must think creatively. Perhaps even more importantly, the analyst's creative abilities are called on not only to work on effective solutions to problems but also to be sure that problems are being correctly identified. Consequently Chapter 11, the second chapter in Part 4, is on creative thinking.

Cost Benefit Analysis

Finally, the analyst must prepare at least two cost benefit analyses, one when the initial feasibility investigation is being made and another before the construction phase is entered. In addition, if the development effort is divided into additional "phases" to implement a creeping commitment approach, the analyst will be called on to prepare more than this minimum number of cost benefit analyses. The procedure for preparing a cost benefit analysis is described in Chapter 12, "Cost Benefit Analysis," the last chapter in Part 4 and in this book.

10

Interpersonal Relations

A meaningful distinction with respect to interpersonal relations is the distinction between *intimate* and *practical* relationships. For example, intimate relationships may exist between spouses, parent and child, siblings, or friends. Typical practical relationships exist between you and your boss, clients, colleagues, and subordinates. The distinguishing characteristic is that in an intimate relationship, the parties to the relationship consider the relationship more important than any goals they may individually or mutually attain by means of the relationship, and participation in the relationship is usually voluntary. In a practical relationship, just the opposite is the case. You enter a practical relationship because you *must* in order to attain a goal; if you consider the relationship practical, this means that the attainment of your goal is more important than the relationship itself. In this book we confine ourselves to considering practical relationships.

To be successful in a practical relationship (that is, to attain the goal that led you to enter the relationship in the first place), you must be *credible*—that is, the other person must believe he understands what both of you are trying to accomplish by means of your relationship. This means you must be honest with the other person with respect to the relationship; you must state what your goal is, why you entered the relationship, what you expect to do in the relationship, and what you expect him to do. And as time passes, you must keep him informed on developments bearing on the relationship.

This credibility requirement simply asks you to be frank about matters that bear on the relationship. It doesn't ask you to reveal intimacies. As a matter of fact, doing so is inimical to a practical relationship. Any proposal of intimacy tends to embarrass the other person and thus encourages him to withdraw from the relationship.

The best way to establish and maintain your credibility in a practical relationship is to keep a clear distinction between your personal self-interest and the attainment of your goal. Be sure that the attainment of the goal is what governs your actions in the relationship; don't let self-aggrandizement enter it. Keep yourself as a person in the background. For example, if your current assignment is to help a client install your company's product, then this goal should dictate all your actions in the relationship. You shouldn't be thinking about how you can impress the other person with your personality, intelligence, and so on. Under such circumstances, the other person will be able to predict your actions, will feel comfortable in the relationship, and won't worry about being surprised by actions on your part that stem from your personal whims.

If we may misuse one of Herzberg's terms, we can say that credibility is the hygenic aspect of practical relationships. If you lack credibility, the other person is discouraged from entering the relationship. However, having credibility isn't enough to encourage the other person to enter the relationship. The thing that will encourage the other person to enter the relationship productively (that is, in such a way that your goal can be attained) is his *self-esteem.*

What we mean by self-esteem is as follows. To survive, every person must have respect for himself; there must be some area in which he feels important. People seek out those situations that give them such a feeling. Discover how a person gets his feeling of importance and you've acquired the key to his actions. Therefore to encourage the other person to enter a practical relationship productively, see that his role in the relationship makes him feel important.

"Fine," you may say, "and how do I go about doing that?" Here are a few suggestions.

1. Ask him to do a favor for you (even a small one) in an area that interests him and about which he knows something. In particular, you're entering a relationship with the other person to attain some goal of yours. Begin the relationship by letting

him know you have a problem you think he can help you solve.

2. Be polite. Phrases such as "I'm sorry to trouble you," "would you be so kind as to," and "thank you" are ways of showing deference to a person you consider important.
3. Use the other person's name when you speak with him. Similarly, find out and discuss with him his interests and achievements. All this shows that you consider him important enough to remember.

Just as there are ways to make the other person feel important, there are way to avoid making him feel unimportant.

1. If you tell jokes, tell them on yourself. Most humor shows something or someone in a ridiculous light. Being made to feel ridiculous is the opposite of being made to feel important.
2. Sometimes the other person's motives aren't ones of which he'd be proud. For example, a client may ask you to help him in making a decision because he wants "an outside opinion"; his real reason may be that he doesn't feel capable of making the decision himself. In such instances the other person may not even suspect his own motives. Don't expose these motives; doing so would damage the other person's self-respect.
3. Don't make moral judgments. A person who does bad things isn't necessarily hopeless. For example, if he's insulting, that doesn't mean you can't deal with him; ignore the insults and get on with the job at hand. If you've approached your relationship with him honestly, your motives must be sound; if you can help him attain the goal of the relationship, you can't avoid improving him. On the other hand, if you condemn him, you'll accomplish nothing: He won't believe you; he won't change his ways; and he won't cooperate with you in the attainment of your goals, either. A good way to get yourself in the proper frame of mind to deal with a person who has faults that offend you is to take a good look at yourself. Whatever fault you find in the other person, be it cruelty, dishonesty, selfishness, or indifference, is it true that you've never acted this way yourself? If you can harbor these traits and still be a good person, why can't the other person do so too? Again, don't attack the other person's self-respect.

If you're about to enter a practical relationship with another person, you're doing so to attain some goal of your own. The first thing you have to do is get the other person's attention. Just as your motive for entering the relationship is to satisfy some need of yours, the other person will enter the relationship only if he feels there's something in it for him. So appeal to him within the context of his experience, interests, and point of view; show him what he stands to gain by cooperating with you.

Once you have the other person's attention, create an atmosphere in which the relationship can grow. Be pleasant and cheerful. Above all, keep smiling. Don't needlessly aggravate the other person; avoid those topics to which he's sensitive. Do small favors for him. And increase his security through appropriate physical contact: the handshake, the pat on the back.

One of the best ways to make a person feel important is to let him do the talking. Therefore, as soon as the other person starts talking, you stop. It doesn't make any difference where you are. You may be in the middle of a sentence, the middle of a word, or at the point of making what you consider to be the most important comment in the whole conversation. Stop talking instantly. The other person wouldn't start talking unless he felt he had something important to say. Once he starts talking, let him have the floor until he's finished. Not only should you let the other person talk, you should encourage him to talk. Ask him questions. In particular, if you want him to feel important, get him talking about himself.

You may be drawn into a practical relationship because another person has a complaint; or during the course of the relationship the other person may bring a complaint to you. When the other person complains, hear him out with respect; any other attitude on your part indicates to him that you don't think he's important enough to have his complaints given due consideration.

The best way to satisfy a complaint is to handle it. However, if it's a complaint that falls outside your domain of concern, action on your part is seldom necessary. Frequently, a person who has a complaint is completely satisfied if he's just heard out with respect and sympathy—it satisfies his need to feel important.

You don't have to agree with a person making a complaint. Just be sure to demonstrate that you understand why he feels the way he does. Show respect for his opinion, and never tell him he's wrong. At the least, you can always say (with complete honesty) that if you were him, you'd react the same way. Listening patiently and sympathetically will often pacify even the most violent critic; anger, unfed by denial, cannot maintain itself.

Then there are arguments. No two people ever agree on everything, so what do you do when you and the other person reach a point of disagreement? The fundamental rule is *never argue.* At best, you'll prove the other person wrong, but even then, he can't admit this without damaging his feeling of importance.

So how do you avoid arguments? To begin with, keep in mind that you entered the relationship in the first place to attain some goal. On every point of disagreement, ask yourself, "Can I yield this point without destroying my chances of attaining my goal?" If you can, concede the point immediately.

If you can't concede the point, see if you can postpone the question. People's attitudes change, and what isn't possible today may be possible tomorrow.

If postponement of the issue isn't possible, then you must discuss it. However, don't ever tell the other person he's wrong. Instead, start by saying, "Well, I thought otherwise, but I may be wrong. Let's look at the facts." Remember that the other person has a reason for his beliefs. While discussing the facts, try to determine these reasons. Find them, and you have the key to changing his attitudes.

Most practical relationships involve selling of some sort. Sometimes you may be selling a product or service. More frequently, what you're selling is an idea. Even when you have a product, the product is really secondary; it's the idea behind the product that needs selling. Once the idea is bought, purchase of the product follows automatically.

The only way to sell another person an idea is to figure out what he wants and show him how your idea can help him get it. If you want him to say "Yes" to your idea, then put him and keep him in a positive frame of mind. Phrase your presentation and questions so the other person is consistently thinking and saying "Yes" from the outset. Don't make your presentation vague; offer a complete, clear-cut program.

One important aspect of selling is knowing when to stop—which is as soon as the other person has bought your idea. To continue to press the advantages of your idea after that is to insult his intelligence. One way you can tell when he's bought your idea is that he'll stop discussing it and start talking about the details of implementing it. You'll recognize this change of perspective only if the other person is talking. So always make your sales presentation a dialogue, not a monologue.

One common selling situation arises when you're in a supervisory position responsible for directing other people's activities. The only way to get someone to do something effectively is to get him to want to do it. The best way to get a person to commit himself to doing something is for him to believe that doing it is his own idea. Therefore, avoid giving orders. Instead, make suggestions, and make them in an unassuming way that defers them to the other person's judgment and thus invites his agreement. Say, "You might consider doing this," or, "Do you think this would work?" or, "What do you think of this?" Let the other person decide that he wants to do the thing.

Convey your suggestions in as general a way as possible. Let the other person work out the details. This approach reinforces the idea that the job is his, not yours. And of course, if the job is his, you must be sure he gets all the credit for its accomplishment. Give proper recognition to even the slightest achievement; this encourages the other person to strive for greater achievement.

It should be clear that when a person makes a commitment to a job, you've given him the responsibility for it. And responsibility means the right to make mistakes. A person learns infinitely more by making a mistake, recognizing it, and correcting it himself than by being corrected by you when he's in the process of making the mistake. Only when it's apparent that he doesn't recognize the mistake should you call attention to it. And when you do, begin by telling him how you've had similar difficulties. This makes him at least as good as you and makes it clear that you don't consider the shortcoming a personal fault.

Life is often rocky, and there will be times when the other person in a practical relationship will criticize your actions. When this happens, take the blame, *even* if you think he's wrong. If you don't, you've violated a basic tenet of conducting a practical relationship—you've chosen to argue on a point not relevent to the attainment of your goal. On the other hand, if you take the blame, the only way left for the other person to feel important is to be magnanimous: He'll belittle the fault and suggest that it be forgotten. You're now ready to resume the pursuit of your goal.

No one is perfect, and on occasion you'll fail to carry out your responsibilities in a practical relationship. When this happens, admit your failure quickly, apologize without qualification, and make a direct appeal to resolve differences. If things are so bad that the other person won't speak to you, appeal through a disinterested party. If the very mention of your name inflames the other person, let some time pass; in time people tend to forget the unpleasantness they've experienced.

Finally, there may be situations in which attainment of your goal is inimical to the attainment of the other person's goal. Under such circumstances, your success requires that you frustrate or even defeat him. However, even in such cases, you should pursue your goal in such a way that you leave the other person with his self-respect intact; after all, you may need his cooperation sometime in the future.

Separate him from the circumstances: Give him reasons for your actions that don't attack his sense of importance; show him that you respect him personally even if you're required to oppose him in a particular situation. Try to suggest some other ways he can attain his goal.

EXERCISES

1. What is the distinction between an intimate and a practical relationship?
2. What is the necessary condition for a successful practical relationship?
3. Once you've established your credibility in the other person's eyes, to what must you appeal to get him to enter a practical relationship productively?
4. List three ways of making another person feel important.
5. List three things you shouldn't do, because doing so makes the other person feel unimportant.
6. If you want another person's cooperation in attaining a goal, how do you get him interested in cooperating with you?
7. List five ways of strengthening a relationship with another person.
8. What should you do when someone brings a complaint to you?
9. Describe a three-step procedure for avoiding arguments.
10. How can you tell when another person has bought your ideas?
11. What should you do when someone criticizes you?
12. Describe a three-step procedure for repairing a relationship damaged by a mistake of yours.
13. If attainment of your goal requires the frustration of another person, name three things you can do to maintain the other person's self-respect even while you're preventing him from attaining his goal.

11.1. DEDUCTION

Let's examine the deduction the merchant's daughter made when she discovered the relevance to her problem of the color of the pebble left in the money bag. It has the following form.

> Premise 1:
> The pebble I drew from the bag and lost on the path must be black or white.
>
> Premise 2:
> It's not black (because the one left in the bag is black).
>
> Conclusion:
> Therefore, it's white.

We don't say the girl actually went through these logical steps. It's almost certain she didn't. But if we were to spell out the thinking she went through almost instantaneously, it would take such a form.

A series of premises followed by a conclusion, such as is shown above, is called an *argument.* Arguments can be *valid* or *invalid,* and an argument's validity or lack of it depends on its form. The example we have here is a valid argument, and it happens to be an argument of the *alternative form.*

Notice that the validity of an argument form doesn't necessarily have anything to do with the truth of either the premises or the conclusion. In the above argument both the second premise and the conclusion happen to be false.

The only relation between *truth* and *validity* is that *if* premises are true, *then* the conclusion of a valid argument is true.

You use deduction for at least three purposes.

1. To determine whether your insights lead to useful conclusions. (This is how the merchant's daughter used deduction.)
2. To determine the implications of given facts. (We'll go into this in more detail in the section on organizing the facts.)
3. To find logical flaws in arguments (yours or other people's). If you can show that an argument is invalid, you cast doubt on the truth of the conclusion without having to determine the truth or falsity of the premises. However, keep in mind that proving that an argument is invalid doesn't prove that the conclusion is false. The conclusion of an invalid argument may still be true; it simply doesn't follow from the premises.

Let's try out our deductive abilities. Look at the following argument. Is it valid? If not, how is it invalid?

> Suppose I am looking at a star, say, Sirius, on a dark night. If physics is to be believed, light waves that started to travel from Sirius many years ago reach the earth (after a specified time, which astronomers can calculate), impinge upon my retinas, and cause me to say I am seeing Sirius. Now the Sirius about which these light waves convey information is the Sirius that existed when they started. However, this Sirius may no longer exist; it may have disappeared in the interval. To say that one can see what no longer exists is absurd. It follows that, whatever it is that I am seeing, it is not Sirius.

Make up your mind as to your answer to this question before reading on.

The answer is that the argument is invalid. Given that it takes some period of time for the light waves that cause sight to travel from the object seen to your eyes, there's no valid way to arrive at the conclusion that you can't see what no longer exists. In fact, the valid conclusion is the opposite one: It *is* possible to see something that no longer exists, despite the fact that this conclusion may be psychologically repugnant to us. Thus, to throw doubt on this conclusion, it's necessary to question the premises from which it flows—the tenets of modern physics.

A description of valid deductive arguments is given in Appendix B.

11.2. TRUTH

Since the truth of our conclusions depends on the truth of the premises from which they've been validly derived, we should talk a little about what establishes the truth of premises. The meaning of "truth" is a dense philosophical question we're going to sidestep in this book. To put it simply and with a large degree of accuracy, truth is what we believe to be true.

Try yourself out on the following exercise.

> True or false: You're never justified in believing a false proposition to be true.

Make up your own mind before reading on.

The proposition is false. Since truth is nothing more than what we believe to be true, a proposition currently believed to be false may, on discovery of further evidence, turn out to be true. This has happened many times in the past and will undoubtedly continue to happen in the future.

Our source of truth is our own *experience*. If we see, hear, or otherwise sense something, we know it's true.

But we can't personally experience everything we need to know. Consequently we rely on *authority*. If an authority says something is true, we accept the statement. For example, ask yourself why you believe that it's the oxygen in the air you breathe which is beneficial to you. Most of us have no evidence for this belief; we accept the statement on authority. As a matter of fact, the great majority of what we believe to be true we believe on the basis of authority rather than on the basis of experience.

11.3. AUTHORITY

If authority is such a pervasive source of truth, then we have to worry about how to tell the authorities from the imposters. Some ways are as follows.

1. *Recognition* of the person as an authority by the community in which he operates. Diplomas and degrees are symbols of such recognition. We tend to give a high degree of credibility to such recognition in the physical sciences, for example, because the practitioners in this field have developed a relatively rigorous and reliable method for granting such recognition.
2. *Confirming opinions.* If the opinions of the purported authorities you consult tend to agree, your confidence in their opinion rises. Thus doctors often call in a consulting physician before deciding on a course of action.
3. *Special competence.* "He was there, so he ought to know." For example, a person who has developed a communications network ought to have some idea of what's involved.
4. *Established credibility.* "He's been right in the past." In ambiguous areas, such as the effect of political decisions or the desirability of an investment, we tend, in the last analysis, to rely on the word of people who have demonstrated their ability to make perceptive judgments.

Test your ability to judge authority with this exercise.

> True or false: A person's statements about his own experience's can't be challenged.

Make up your mind before reading on.

The statement is false. The person has special competence on his side, but his credibility may still be poor.

Here's another exercise on judging authority.

> Your friendly hardware salesman tells you you'll receive delivery of your equipment two weeks from today. Should you believe him? What claims to authority can he make?

First of all, your salesman has *recognition* on his side. He's the hardware manufacturer's accredited representative.

You can get confirming opinions from your salesman's manager and his company's production department. If your salesman didn't make up the delivery date, these people will *confirm* what he says.

However, your salesman doesn't have *special competence.* You'd be better off checking with other people who have taken delivery from him before you rely on his statement.

Finally, it might pay you to check into the *established credibility* of the authorities on whose statements you may plan to rely. Has your salesman, his manager, and his company's production department been wrong before? Under what circumstances have they been right? Under what circumstances have they been wrong?

Here's one final exercise on judging authority.

> The vice-president of a division of your company tells you that each of his department heads makes extensive use of a particular report. Should you believe him?

Strange as it may seem at first blush, your vice-president has about the same qualifications as an authority as does your hardware salesman.

1. He's a recognized authority.
2. You can get confirming opinions from persons performing similar operations in other divisions of your company or in similar divisions of other companies. There may even be literature (articles and books) on the subject.
3. Your vice-president doesn't have special competence. The people who are where the action is are the department heads.
4. You must judge your vice-president's established credibility. Does he usually know what he's talking about?

11.4. INDUCTION

As we've said, induction is concerned with gaining insight from collected information. At least five types of activity can be recognized as parts of the inductive process.

1. Stating the problem
2. Collecting the facts
3. Organizing the facts
4. Developing ideas
5. Testing the ideas

There's some tendency to say that these activities occur in the sequence listed, and in broad terms this may be the case. However, there's considerable overlap and leaping from one activity to another.

1. It's unlikely that even the attempt to state a problem will be made without some facts at hand.
2. The way the problem is stated influences the selection of the facts and the way they're organized. At some point the increasing sophistication acquired by trying to organize the collected facts satisfactorily may lead to a restatement of the problem.
3. Ideas can occur anytime. An unexpected but fruitful idea may feed back on the way the facts are organized and even modify the problem statement.
4. Testing ideas is also a continuous process. Many tests are purely logical in nature and are applied as soon as the idea occurs. Others call for the collection and consequent organization of facts previously considered irrelevant. Some test results suggest further ideas, and the feedback cycle begins again.

As deBono says: "In practice, thinking is a messy business" [3].

With this caveat out of the way, let's take an individual look at each of the activities making up the inductive process.

11.4.1. Stating the Problem

It should go without saying that precise statement of a problem is a long step toward solving the problem. After all, it's pretty hard to get someplace if you haven't spelled out where you're going. Nevertheless, one of the more common causes of failure in problem solving is insufficient time spent in coming up with a precise problem definition.

In their book *The Rational Manager* [5], Charles Kepner and Benjamin Tregoe recount a story that exemplifies the importance of precise problem statements. A paper manufacturer had a pulping plant connected to a paper machine. In the pulping plant softwood logs were converted to pulp, which was then strained and fed into the paper machine via a pipe. One day splinters were found in the paper being produced. It was assumed that a screen used to strain the pulp had broken, and $70,000 of new equipment was ordered. However, one man examined the splinters and found that they were hardwood, not softwood. Armed with this information, he then discovered that the pipe connecting the pulping plant to the paper machine was made of hardwood. As a result, he hypothesized that the pipe was breaking up on the inside. Investigation proved him right. Ordering unnecessary new equipment could have been avoided if the problem had been defined at the outset as "*hardwood* splinters in the paper" rather than just as "splinters in the paper."

Data processing system development and modification seem to be particularly susceptible to inadequacies in problem statement. Most system development and modification begin with a problem statement, and the problem statement is often provided by the user as part of his request for services. In a course I teach on system development methodology, I introduce the topic of problem definition by suggesting that the attendees and I play the following game. They're system analysts and I'm a user. Our organization has an order entry system to which I want a modification made. I state the problem as follows.

> I need a report that will keep me informed on the aging of back orders.

I then ask the attendees to respond to this request as system analysts. Invariably this leads to a discussion of the format, sequencing, and frequency of the requested report. Often the discussion ends right there. Sometimes it dawns on an attendee that what I've described is a solution, not a problem. Then they ask the appropriate question: "What are you going to use the report for?"

In answer to this question I tell them that I'm going to use it so I can see that back orders get filled on a first-in, first-out, basis—that is, when stock is replenished, the oldest back orders get filled first. On the occasions where the conversation has gotten this far, the class attendees invariably accept this answer, and the discussion ends. No attendee has ever recognized that my answer is another solution statement and that the problem remains undefined.

If they were to ask me why I want to fill the back orders on a first-in, first-out basis, I'd say that the longer an item stays on back order, the more likely the order is to be canceled. Only now is the problem beginning to be uncovered. Order cancellations must be undesirably high, or I never would have developed a concern for getting the oldest back orders off the books first, and I'd never have asked for the back order report in the first place. The problem is too high a rate of order cancellation.

This example also emphasizes why good problem definition is important. Once we determine that it's the cancellation rate which lies at the root of the problem, we can see that better inventory or production control techniques may be a more effective solution to the problem than any monitoring of back orders.

A good problem statement begins with two things.

1. A description of the present situation
2. A standard of desired performance

Problems are then stated in terms of how the present situation deviates from the standard of performance. These deviations must be precisely defined—what is deviating, where the deviation is occurring, when it's occurring, and the extent to which it occurs.

The beginning of a good problem statement might be as follows.

> The longer an item is on back order, the more likely cancellation becomes.

Here the standard of performance is no cancellations (or some reasonable approximation of this), and the problem statement describes when and to what extent deviations from this standard occur. (Of course, we are assuming that this kind of "headline" statement of the problem can be backed up by statistics demonstrating quantitatively the increase in occurrence of cancellations as length of time on back order increases.) Now effort can be productively spent in determining what circumstances create back orders; to what extent these circumstances result in back orders; what influence elimination of these circumstances would have on order cancellation; what is required to eliminate these circumstances; whether, in fact, any solutions to this problem lie in the data processing area; and what these solutions might be.

Try the following problem-solving exercise.

> During World War II, a bomber's hydraulic control system was damaged by antiaircraft fire. On the flight home, the crew patched up the system, but they couldn't find any fluid to top off the system. How did they save themselves?

The crew saved themselves by topping off the system with urine. This exercise is an example of a problem that's more readily solved the more precisely it's defined. What was needed was a fluid—any fluid. Only when this was recognized might urine and hydraulic systems be thought of in the same context.

Kepner and Tregoe's book *The Rational Manager* is an extensive investigation into the nature of problem definition and how it's used in problem solution. We can't recommend this book too highly.

11.4.2. Collecting the Facts

Since the problem statement will guide you in deciding what facts are pertinent and thus what facts should be collected, the importance of taking the time to come up with a precise problem statement before formal fact collection must once more be emphasized.

11.4.3. Organizing the Facts

The more facts you have bearing on your problem, the more likely you are to come up with an effective solution. However, the pertinence of information is often hidden by the quantity of information available. Organizing the information is a way of causing the pertinence to become evident. For example, the organization of the chemical elements in the periodic table led to the discovery of pertinent relationships involving the characteristics of the chemical elements.

On a more rudimentary level, the contribution that organization of facts can make to problem solution is well exemplified by problems such as the following, which is taken from Bennett and Baylis, *Formal Logic,* as quoted in Black, *Critical Thinking* [1].

> Smith and Brown each won $10 playing poker with the pitcher. Hunter is taller than Knight and shorter than White, but each of these weighs more than the third baseman. The third baseman lives across the corridor from Jones in the same apartment house. Miller and the outfielders play bridge in their spare time. White, Miller, Brown, the right fielder, and the center fielder are bachelors. The rest are married. Of Adams and Knight, one plays outfield. The right fielder is shorter than the center fielder. The third baseman is the pitcher's wife's brother. Green is taller than the infielders and the battery, except for Jones, Smith, and Adams. The second baseman, the shortstop, and Hunter made $150 each speculating in U.S. Steel. The second baseman is engaged to Miller's sister. Adams lives with his own sister but dislikes the catcher. Adams, Brown, and the shortstop lost $200 each speculating in copper. The catcher has three daughters, the third baseman has two sons, but Green is being sued for divorce. Who plays what positions?

Here you're given all the information needed to solve the problem. Yet you could probably stare at this information in its present form forever without arriving at the solution. However, if you methodically go about organizing the information, you can arrive at the solution with no creative thought whatever. All that's needed is some deductive reasoning. For example, we're told that Green is taller than the infielders and the battery, except for Jones, Smith, and Adams. This tells us that Adams is either an infielder or a member of the battery (premise 1). We also know that there are only three types of players: infielders, the battery, and outfielders (premise 2). Therefore (conclusion) Adams is not an outfielder. (This conclusion is also premise 1 for the next argument). We're also told that, of Adams and Knight, one is an outfielder (premise 2). Therefore, Knight is an outfielder (conclusion). And so on. (Both of the example arguments given here are of the alternative form.)

There are two activities in classifying information.

1. Setting up a classification scheme
2. Sorting the information into the scheme, which generally requires the use of deduction

Again, there's a tendency to consider this a two-step process, but in reality there's considerable interrelation between the two activities. The first classification scheme is usually preliminary, and to contain the collected information satisfactorily, it requires continuous modification as the sorting proceeds.

Here's a problem-solving exercise involving fact organization.

> Potter has never been married. Barnes and Carter are brothers-in-law. Turnquist and O'Toole have the same marital status. Two of the five men are Californians, and three are Texans. None of the men has ever been divorced; one is a widower; only two (both Texans) have never been married. If Barnes is a widower, what is the marital status of Carter?

We'll use a grid on which the given information can be organized.

Potter has never been married (Figure 11.1).

Barnes is a widower; moreover, only one is a widower (Figure 11.2).

Turnquist and O'Toole have the same marital status. This status can't be "never married," because only two men have never been married, and Potter is one of them. Therefore both Turnquist and O'Toole are married (Figure 11.3).

	NEVER MARRIED	WIDOWER	MARRIED
POTTER	X		
BARNES			
CARTER			
TURNQUIST			
O'TOOLE			

Figure 11.1

	NEVER MARRIED	WIDOWER	MARRIED
POTTER	X		
BARNES		X	
CARTER			
TURNQUIST			
O'TOOLE			

Figure 11.2

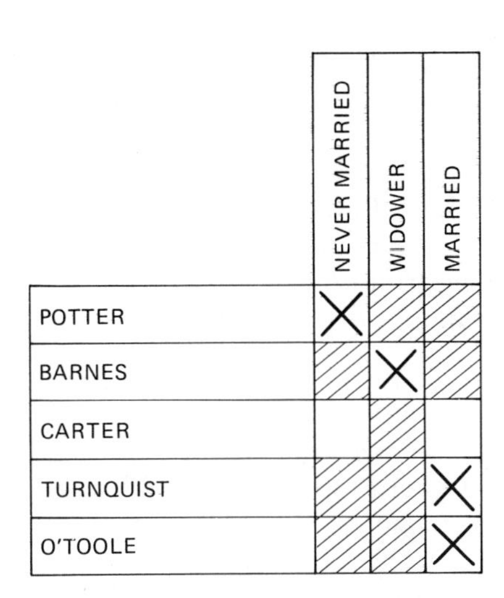

Figure 11.3

Two men have never married. Potter is one of them. Carter must be the other. Therefore Carter has never been married.

Notice that the state in which the men live and the fact that Barnes and Carter are brothers-in-law are irrelevant to the solution. It's common in problem-solving situations to discover that information collected ultimately turns out to be irrelevant. This shouldn't worry you; you should consider it nothing more than an occupational hazard.

Sometimes organization of information alone isn't enough to solve the problem. Assumptions must also be made. For example, consider the following problem.

1. There are five houses.
2. The Englishman lives in the red house.
3. Milk is drunk in the middle house.
4. The Spaniard owns a dog.
5. The Japanese smokes Parliament.
6. The Norwegian lives in the first house.
7. Kools are smoked in the house next to where the horse is kept.
8. The man who smokes Chesterfield lives next door to where the fox is kept.
9. The man who smokes Old Gold owns snails.
10. The man who smokes Lucky Strike drinks orange juice.
11. The green house is immediately to the right of the ivory house.
12. The man in the yellow house smokes Kools.
13. The man in the yellow house drinks coffee.
14. The Norwegian lives next door to the blue house.
15. The Ukranian drinks tea.
16. Each man has a different house, drink, smoke, pet, and nationality.

Who drinks water? Who owns the zebra?

A possible approach to solving this problem is as follows.

There are six characteristics given: house order, nationality, house color, drink, pet, and smoke. With the exception of house order, these characteristics are listed below in the order they appear in the problem statement.

Englishman	Red	Milk
Spaniard	Green	Orange juice
Japanese	Ivory	Coffee

Norwegian	Yellow	Tea
Ukranian	Blue	Water

Dog	Parliament
Horse	Kools
Fox	Chesterfield
Snails	Old Gold
Zebra	Lucky Strike

We're given certain information about house order.

> Milk is drunk in the middle house.
>
> The Norwegian lives in the first house.
>
> The green house is immediately to the right of the ivory house.

Since there's a first house, the houses must be arranged in a line rather than a circle. It's customary to number things in a line from left to right, so let's assume that the person who stated the problem followed this custom. (Notice that this an assumption; there's nothing in the problem statement to guarantee this numbering method.) Let's number the houses 1, 2, 3, 4, and 5. Then the Norwegian lives in house 1, milk is drunk in house 3, and the green house number must be one larger than the ivory house number. Let's indicate these facts in two ways.

1. If we know the house number of a characteristic, we'll put it in parentheses following the characteristic in our list of characteristics.
2. We'll put our lists in numerical order from top to bottom and from right to left.

Thus, our lists now look as follows.

Nor (1)	Red	Cof
Span	Ivor	OJ
Jap	Green	Milk (3)
Eng	Yel	Tea
Uk	Blue	Wat

Dog	Parl
Horse	Kool
Fox	Chest
Snail	Old
Zeb	Luck

Notice that we've abbreviated many of the characteristics names to make writing the lists easier. We'll also adopt another kind of shorthand and call characteristics by their number. Thus we'll say, "Nor is 1," "Milk is 3," and so on.

We're also given certain information about relations of characteristics, such as:

> The Englishman lives in the red house.
>
> The Spaniard owns a dog.

In our abbreviated language we'll describe these relations as follows.

Eng is Red.
Span is Dog.
Jap is Parl.
Old is Snail.
Lucky is OJ.
Kool is Yel.
Cof is Yel.
Uk is Tea.

Figure 11.4 shows a grid on which we can record these relations. This grid will be helpful in organizing our information, because when we have all the characteristics listed in house number order and all the relations between characteristics indicated, each matrix in the grid will have a diagonal of X's running from the top left-hand corner to the lower right-hand corner. Consequently if an X appears in any other place in the grid as we work toward the solution, we'll know that some list of characteristics is not yet in house number order.

Now let's see what further relations we can deduce from those we already know.

1. Nor lives next door to Blue.
2. Nor is 1.
3. Therefore Blue is 2.

1. Nor is 1.
2. Therefore Eng isn't 1.
3. Red is Eng.
4. Therefore Red isn't 1.
5. Blue is 2.
6. Ivor and Green are next to each other.
7. Therefore Ivor and Green aren't 1 and 2.
8. Therefore Yel is 1.

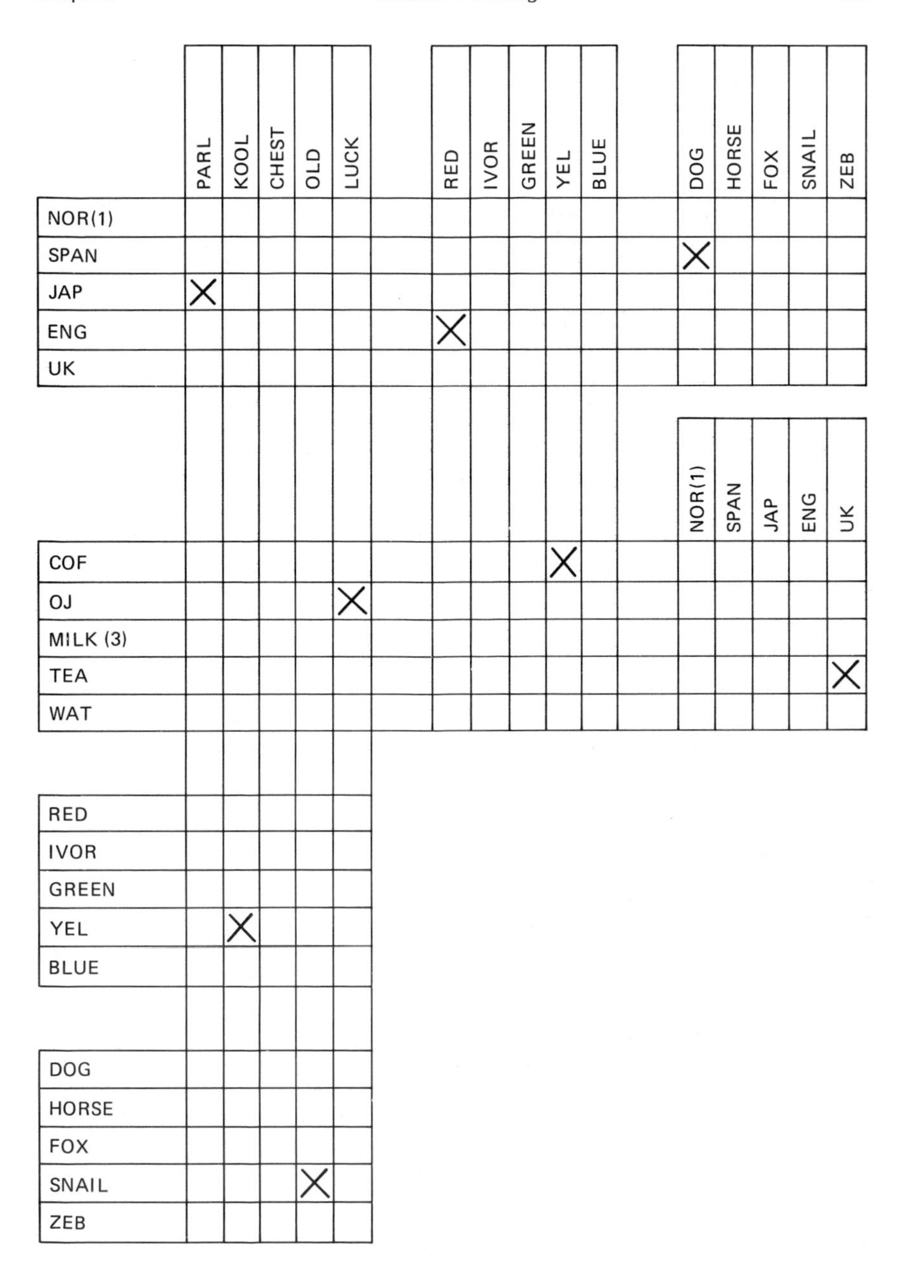

Figure 11.4

1. Yel is 1.
2. Kool is Yel.
3. Therefore Kool is 1.

1. Yel is 1.
2. Cof is Yel.
3. Therefore Cof is 1.

1. Kool is 1.
2. Horse is next to Kool.
3. Therefore Horse is 2.

Another deduction we can make is as follows.

1. Milk is 3.
2. Luck is OJ.
3. Therefore Luck isn't 3.

We'll write "not 3" as "-3." Thus we'll write, "Luck (-3)." Some other deductions we can make are as follows.

1. Blue is 2.
2. Eng is Red.
3. Therefore Eng isn't 2.

1. Nor is 1.
2. Dog is Span.
3. Therefore Dog isn't 1.

1. Horse is 2.
2. Span is Dog.
3. Therefore Span isn't 2.

1. Kool is 1.
2. Snail is Old.
3. Therefore Snail isn't 1.

1. Ivor and Green are together.
2. Therefore they must be 3 and 4 or 4 and 5.
3. Therefore Red isn't 4.

1. Red isn't 4.
2. Eng is Red.
3. Therefore, Eng isn't 4.

1. Luck isn't 3.
2. OJ is Luck.
3. Therefore OJ isn't 3.

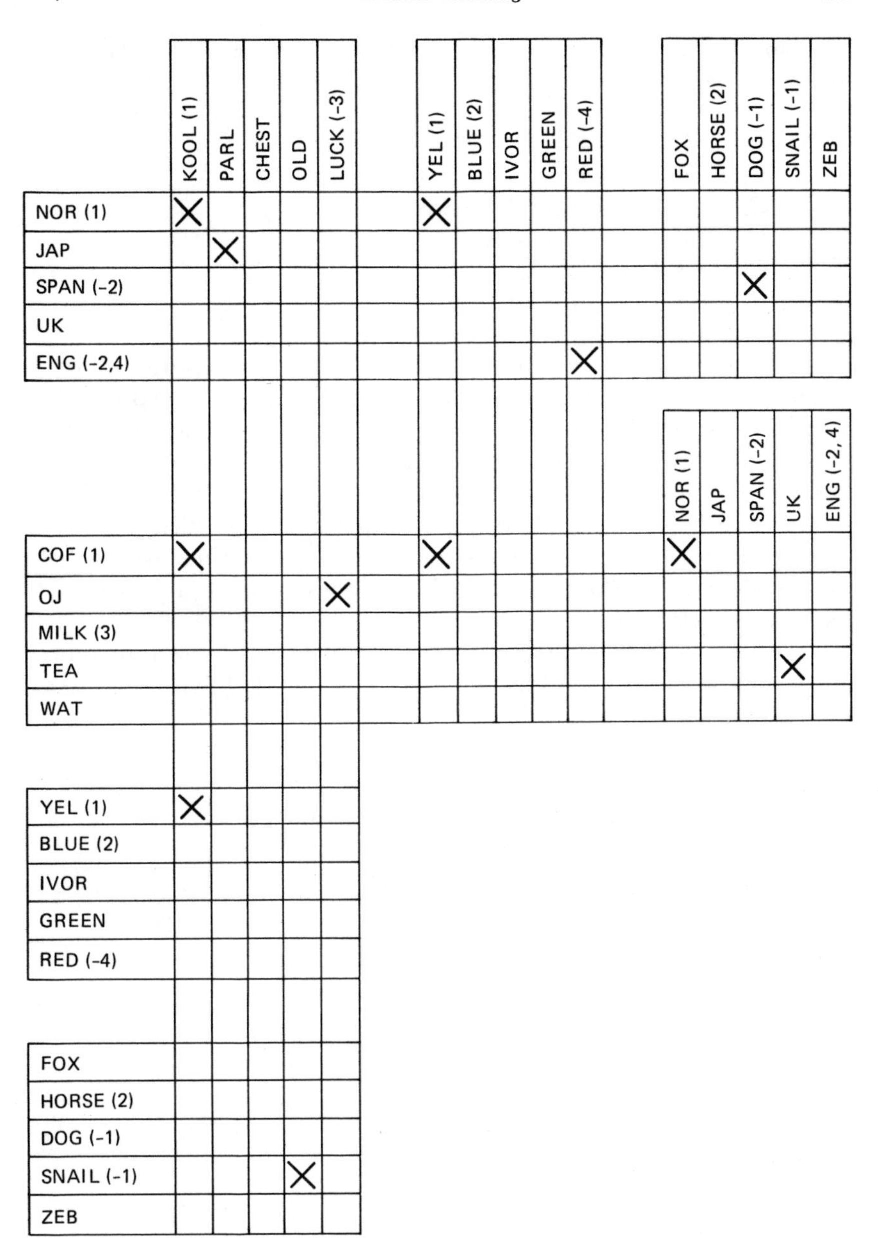

Figure 11.5

The given facts and the facts that can be directly deduced from them are shown in Figure 11.5. To go further, we have to start adopting hypotheses.

One fact we haven't used is that Chest is next to Fox. This could mean any of the following.

1. Chest is 2 and Fox is 1.
2. Chest is 2 and Fox is 3.
3. Chest is 3 (and therefore Fox is 4).
4. Chest is 4 and Fox is 3.
5. Chest is 4 and Fox is 5.
6. Chest is 5 (and therefore Fox is 4).

We'll have to adopt one of these possibilities as a hypothesis and see where it leads us. The choice is arbitrary, since any of the possibilities could be the correct hypothesis, but as long as we have the choice, we ought to choose one that looks as if it will lead to fairly direct results. On this basis, possibility (1) is a poor choice. Of all the possible hypotheses, only (1) doesn't immediately establish Nor as the owner of the zebra. By the same kind of reasoning, we probably shouldn't choose Chest as 3, since this doesn't make the most of the fact that Luck isn't 3.

There's little difference in the attractiveness of the four remaining possibilities. So let's start at the top and work down. Let's assume that Chest is 2 and Fox is 3.

We can now make the following deductions.

1. Fox is 3.
2. Dog and Snail aren't 1.
3. Therefore Zebra is 1.

1. Fox is 3.
2. Span is Dog.
3. Therefore Span isn't 3.

1. Chest is 2.
2. Jap is Parl.
3. Therefore Jap isn't 2.

1. Jap, Span, and Eng aren't 2.
2. Therefore Uk is 2.

1. Uk is 2.
2. Tea is Uk.
3. Therefore Tea is 2.

This information is shown in Figure 11.6.

	KOOL (1)	CHEST (2)	PARL	OLD	LUCK (-3	YEL (1)	BLUE (2)	IVOR	GREEN	RED (-4)	ZEB (1)	HORSE (2)	FOX (3)	SNAIL (-1)	DOG (-1)
NOR (1)	X					X					X				
UK (2)		X					X					X			
JAP (-2)			X												
SPAN (-2, 3)															X
ENG (-2, 4)										X					

	KOOL (1)	CHEST (2)	PARL	OLD	LUCK (-3	YEL (1)	BLUE (2)	IVOR	GREEN	RED (-4)	NOR (1)	UK (2)	JAP (-2)	SPAN (-2, 3)	ENG (-2, 4
COF (1)	X					X					X				
TEA (2)		X					X					X			
MILK (3)															
OJ					X										
WAT															

	KOOL (1)	CHEST (2)	PARL	OLD	LUCK (-3
YEL (1)	X				
BLUE (2)		X			
IVOR					
GREEN					
RED (-4)					
ZEB (1)	X				
HORSE (2)		X			
FOX (3)					
SNAIL (-1)				X	
DOG (-1)					

Figure 11.6

This is as far as our assumption concerning Chest and Fox will take us. To go further, we'll have to make another assumption.

Following our rule that we should make assumptions that lead to fairly direct results, we can zero in on a hypothesis as follows. We're currently operating under the assumption that Nor owns the zebra. We now want to find out who drinks the water. So we ought to make an assumption about drinks. There are two possibilities.

1. Wat is 4.
2. Wat is 5.

There appears to be little to recommend either hypothesis. So let's start at the top and assume that Wat is 4. We can now make the following deductions.

1. Wat is 4.
2. Therefore OJ is 5.

1. OJ is 5.
2. Luck is OJ.
3. Therefore Luck is 5.

1. Luck is 5.
2. Snail is Old.
3. Therefore Snail is 4.

1. Snail is 4.
2. Old is Snail.
3. Therefore Old is 4.

1. Therefore Parl is 3.

1. Parl is 3.
2. Jap is Parl.
3. Therefore Jap is 3.

1. Jap is 3.
2. Eng isn't 4.
3. Therefore Span is 4.

1. Span is 4.
2. Dog is Span.
3. Therefore Dog is 4.

1. But Snail is 4.
2. Therefore if Chest is 2 and Fox is 3, Wat can't be 4.

This is an example of a valid argument type called *reductio ad absurdum.*

We're left with the remaining possibility concerning water. So let's assume that Wat is 5. We can now make the following deductions.

1. Wat is 5.
2. Therefore OJ is 4.

1. OJ is 4.
2. Luck is OJ.
3. Therefore Luck is 4.

1. Luck is 4.
2. Snail is Old.
3. Therefore Snail is 5.

1. Therefore Dog is 4.

1. Dog is 4.
2. Span is Dog.
3. Therefore Span is 4.

1. Snail is 5.
2. Old is Snail.
3. Therefore Old is 5.

1. Therefore Parl is 3.

1. Parl is 3.
2. Jap is Parl.
3. Therefore Jap is 3.

1. Therefore Eng is 5.

1. Eng is 5.
2. Red is Eng.
3. Therefore Red is 5.

1. Therefore Ivor is 3 and Green is 4.

These conclusions lead to the following lists of characteristics.

1. Nor	1. Yel	1. Cof
2. Uk	2. Blue	2. Tea
3. Jap	3. Ivor	3. Milk
4. Span	4. Green	4. OJ
5. Eng	5. Red	5. Wat

1. Zeb	1. Kool
2. Horse	2. Chest
3. Fox	3. Parl
4. Dog	4. Luck
5. Snail	5. Old

As this listing indicates, it's possible for the Norwegian to own the zebra and for the Englishman to drink water.

The only remaining question is: Is this solution unique? There are five other possibilities involving Chest and Fox that haven't been investigated. You can show using the *reductio ad absurdum* form of argument that each of these remaining possibilities leads to inconsistent conclusions. Once this is demonstrated, it can be stated with certainty that it's the Norwegian who owns the zebra and the Englishman who drinks the water.

11.4.4. Developing Ideas

Up to now we've confined ourselves to problems which, if correctly stated, allow us to deduce the solution from the collected information. However, many problems can be solved only after it is recognized that certain information has a *nondeductive* relationship to the solution. This is the heart of the inductive process: the insight into the relevance of information to the problem at hand. For example, the merchant's daughter experienced insight when she recognized the relevance to her problem of the color of the pebble remaining in the bag.

Experiencing insight is commonly referred to as creative thinking. Let's look at some of the efforts that have been made to understand the creative process and see what these investigators have found.

11.4.4.1. What Is Creative Thinking? Most authorities agree that a creative thought must have two characteristics: It must be both *new* and *practical.*

By "new" we mean that the creative thought must be, to the person having the thought, an original organization of information, an arrangement the person isn't aware of anyone having come up with before. This doesn't mean the thought has to be totally new, but it must have some aspect of originality. For example, keeping tennis balls under pressure to preserve their bounce isn't a new idea; tennis balls have been packaged this way for a long while. But selling a permanent can with a built-in hand pump so you can keep your used tennis balls under pressure when you're not playing with them is a new idea.

We must also bear in mind that originality is strictly a personal thing. We don't deny that an idea is original with one person just because someone else has already thought of it. Thus the concurrent, independent invention of calculus by Newton and Leibniz was new to each of them, but only because each was unaware of the other's thoughts.

When we say a creative thought must be practical, we mean that the thought must have a productive application. Creative thinking is used in so-called problem areas. A problem may arise in the form of some undesirable situation. For example, skis don't grip slopes covered with hardpack; this problem is solved by edging skis with metal. Or we may say that a problem exists when someone recognizes that a generallly satisfactory situation can be improved. Developing a better water faucet is an example. Most applied engineering falls into this category, as does revision or creation of a data processing system. A special case of "problem solving" is scientific inquiry, where there may be no "problem to be solved" in the sense of a situation requiring corrective action, but only an attempt to explain natural phenomena satisfactorily.

We can refer to all these activities as *problem-solving activities.* Then we can say that, to be practical, an idea must actually solve problems.

A book entitled *The Creative Organization,* edited by Gary Steiner [10], records the transcript of a seminar on this subject. In this seminar, Jerome Bruner summarized the two essential characteristics of a creative thought in the concept of "effective surprise." An effective surprise is something that's surprising (new) and yet effective (the appropriate thing). Mr. Bruner's point is that a creative idea is one which is obvious once thought of and yet surprisingly new when first proposed. For example, the solution the merchant's daughter applied to her problem exhibits the characteristic of effective surprise. Once you see the solution to the problem, it's appropriateness is so compelling that it becomes the obvious choice for a course of action. But when it's first proposed, it comes as a surprise. Thus the idea is both effective and a surprise.

11.4.4.2. Characteristics of Creative People. One approach to the study of creativity has its origins in the observation that some people are more creative than others. The question then posed is, What characteristics distinguish creative people from others?

Here's a list of the traits these investigators have isolated.

1. *Sensitivity:* the ability to see problems
2. *Evaluation:* the ability to distinguish between ideas that will and ideas that won't solve problems
3. *Fluency:* the ability to generate a large number of ideas rapidly
4. *Flexibility:* the ability to change frames of reference
5. *Elaboration:* the ability to embellish an idea with details
6. *Transformation:* the ability to combine things in new ways
7. *Originality:* the ability to come up with unusual (that is, statistically infrequent) ideas
8. *Preference for complexity*
9. *Independence of judgment:* willingness to hold a position in the face of disagreement
10. *Separation of information from source:* the tendency to evaluate information by itself rather than on the basis of its source
11. *Relativism:* the tendency to look at things in relative, rather than absolute, terms
12. *Good sense of humor*
13. *Rich fantasy life*
14. *Deviance:* The characteristic of seeing oneself as different from others

Some investigators have become fascinated by the ways in which the characteristics of creative people seem to be diametric opposites. Bruner calls them "antimonies," and he claims to have identified four. They all seem similar to me, the thrust being that the creative person is both a highly committed, passionate person intent on making decisions and a detached, reflective person with a tendency to defer decisions. Some people say a creative person is both more prejudiced and more open-minded than others. William Schockley (quoted in Steiner) says the creative person is:

1. Both more primitive and more cultured
2. Both more destructive and more constructive
3. Both crazier and saner

than the average person. More objectively, personality tests show more creative persons combining schizoid tendencies (bizarre, unusual, unrealistic thoughts and urges) with unusual ego strength (ability to control, channel, and manipulate reality effectively).

11.4.4.3. Factors Inhibiting Creativity. Another approach investigators of creative thinking have taken is the pathological one. The question they ask is, What keeps a person from being creative? Here's a list of factors I've compiled from the literature, primarily from Steiner.

1. Insufficient intelligence
2. Too schizoid
3. Too much intrusion of fantasy
4. Fear of criticism
5. Excessive anxiety
6. Too low self-esteem
7. Compulsiveness
8. Preoccupation with personal problems
9. Commitment to the status quo
10. Premature commitment to a given line of thought
11. Exaggeration of the value of information that looks potentially useful
12. Inability to see problems
13. Too much involvement in helping other people

It's interesting that some of these factors also appear on the list of characteristics of creative people. For example, having a rich fantasy life is a characteristic of creative people, but apparently you can have too much of a good thing; too much fantasy can choke off creativity.

11.4.4.4. Deferment of Judgment. One of the pioneers in the investigation into creativity is Alex Osborn, cofounder of the advertising firm of Batten, Barton, Dorstine, and Osborn. Mr. Osborn wrote the book *Applied Imagination* [7], founded the Creative Education Foundation, and created the concept of *brainstorming.*

In a brainstorming sessions the object is to come up with ideas without regard to their practicality or impracticality. Criticism of any idea, yours or somebody else's, is forbidden; all judgment is deferred until the end of the session.

The deferment of judgment practiced in a brainstorming session allows many apparently harebrained ideas, which would never by themselves survive the onslaught of critical review, to survive long enough to be modified and combined with other ideas. This results in practical solutions that otherwise would probably never be thought of.

In his *Creative Behavior Guidebook* [8], Sidney Parnes endorses Osborne's principle of deferment of judgment and adds to it his own principle of "extended effort." Here the idea is that, after you've come up with all the ideas you can with respect to solving a problem, you're asked to extend your effort and come up with even more ideas. Studies indicate that the ideas developed as a result of extended effort tend to be more effective problem solutions than the ideas produced earlier.

These are the facts. Now, how do they fit into a consistent theory of creative thinking?

11.4.4.5. Synthesis. Now that we have collected our facts, we must separate the significant from the irrelevant before we can make further progress in understanding. Let's first look at our characteristics of creative people.

In fact, the first seven qualities listed aren't really characteristics of creative people at all. Rather, they're aspects of our original definition of creativity. Sensitivity and evaluation have to do with the practical aspects of a creative thought; fluency, flexibility, elaboration, transformation, and originality have to do with the novelty of a creative thought.

To clarify this point, let us again consider the story of the merchant, his daughter, and the money lender. The daughter arrived at the resolution of her dilemma by applying the principle of flexibility. In some way, she was able to take her mind off the terrible implications of the color of the pebble she had to select, and to change her frame of reference to the color of the pebble left in the bag. Once she had made this switch, she was well on the way to solving her problem. (It seems to me that it's this ability to change frames of reference which provides the surprise element in Bruner's concept of effective surprise.)

Now, let's look again at the words we used in the preceding paragraph to describe the creative insight. When we do, we can see that flexibility isn't a *characteristic* of a creative person; instead, it's a principle, a technique, a method people use to come up with creative thoughts. We started referring to characteristics of creative people when we said that "in some way" the girl was able to apply the principle of flexibility. The question we're trying to answer is, What characteristics does a person need in order to make effective use of the principles of flexibility, elaboration, transformation, and so on.

It's the last seven items in our list of characteristics of creative people that are true characteristics. It isn't until we get to these last seven that we stop saying that creative people must have practical and novel ideas and begin enumerating characteristics of people who have such ideas. Consider these last seven characteristics again:

Preference for complexity
Independence of judgment
Separation of information from source
Relativism
Good sense of humor
Rich fantasy life
Deviance

Is there any common thread that runs through all these characteristics? I believe there is, and the name I'd give it is *nonconformity*. It seems that the creative person tends not to conform to a generally expected behavior pattern.

Now, on to what Bruner calls antimonies. As we've indicated, there are many different ways to express these antimonies. But as we've also implied, Bruner's four antimonies appear to be different aspects of the same thing; thus it seems that there's only one fundamental antimony. This fundamental antimony is probably best expressed by the results of the personality tests we mentioned: Creative people combine schizoid tendencies with great ego strength. But while we're now using high-powered psychological terms, are we really saying anything other than that creative people are those who have ideas which are both new (schizoid) and practical (manipulate reality effectively)? I think not, and if I'm right, then the concept of antimony doesn't help in understanding creative thinking. We can agree that the creative person is one who allows his imagination free range even while testing the products of his free-ranging imagination against the practicalities of solving the problem at hand, but this doesn't add to our knowledge of creative thinking; it's just a restatement of the definition of creative thinking.

Now let's turn to an examination of factors inhibiting creativity. Once again, it seems that if we're going to make any sense of the list of characteristics, we have to separate the pertinent factors from the irrelevant ones. Here goes.

We can concede that a person who has insufficient intelligence won't come up with creative ideas. But this does nothing to further our understanding of creativity, because it leaves us with the question of why so many people of at least average intelligence aren't more creative.

Similarly, it may be true that some people aren't creative because they lack control of their imaginations, but this observation isn't very helpful. Thus when investigators say that some people aren't creative because they're too schizoid or because they allow the intrusion of too much fantasy material, all they're really saying is that possession of a free-ranging imagination is a necessary but not sufficient condition for creative thinking. The imagination must be controlled and not allowed to run wild. But once more, this is just the definition of creative thinking; it doesn't tell us why the average person isn't more creative.

Once we get past these "factors," which are actually just aspects of the definition of creativity, our list becomes more useful. People who fear criticism, who are anxious, who don't have enough self-esteem, who are compulsive, and who are preoccupied with personal problems are *insecure* people. Note the difference between an insecure person (noncreative) and a nonconformist (creative). Both feel that they aren't conforming, but the insecure person wants to conform and worries about it, while the nonconformist is secure in his nonconformity.

And what are the outward characteristics of insecure persons? Unless they break down (which makes them abnormal), they give the appearance of extreme conformity. They're strongly committed to the status quo, and this leads to all sorts of things: They're prematurely committed to lines of thought that conform to the status quo; they select and exaggerate the importance of information that fits in with their hastily adopted ideas, to the exclusion of other potentially pertinent information; and as a result, they don't see many problems.

This leaves just one factor that we haven't discussed: overinvolvement with helping other people. This is the so-called social worker syndrome, or uplift pattern, and according to Paul Meehl (quoted in Steiner), it's strongly and inversely correlated with creativity. This seems logical, because the overinvolvement may be just one more mask behind which the insecure person can hide. After all, concentrating on the other person's inadequacies may be a way of getting your mind off your own.

So it seems there's also a common thread running through the factors inhibiting creativity. The name to give this common characteristic is *insecurity*, which we might most effectively define as fear of nonconformity.

We've now come quite far in our attempt to develop a theory of creative thinking. We've said that the creative thinker tends to be a nonconformist, and we've just reinforced this idea by discovering that the basic influence inhibiting creativity is insecurity—fear of being labeled a nonconformist. Let's see if we can use this theory to explain the success of brainstorming and extended effort in improving the generation of creative ideas.

The relationship between our theory and the success of brainstorming is made evident by Osborn's statement of the principle of deferment of judgment. If all criticism of ideas is taboo, then at least the external and explicit condemnation of "weird" ideas as evidence of nonconformity has been removed, the security of the participants in the brainstorming session is increased, and creativity increases in proportion.

The relationship between our theory and the principle of extended effort may not be so evident, but fortunately, Parnes makes it clear. Before extended effort begins, Parnes explains, a person gets the opportunity to exhaust his store of "safe" (conforming) ideas. Then when extended effort begins, the person has no alternative but to start generating "unsafe" ideas. So once more, we see that the principle of extended effort is no more than a mechanism, similar to the mechanism of deferment of judgment, for helping a person overcome insecurity and start generating what might otherwise be thought of as nonconforming ideas.

To be sure, the ideal approach to creative thinking is to avoid generating inhibitions, but I'm skeptical of our ability to make any significant inroads in this direction in the near future. There is extensive documentation of how our school system encourages conformity and squashes any sign of independent thought. And things are no better in industry. For example, did you know that frequency of contribution to a company suggestion plan is inversely related to rate of promotion within the company?

So given our present dismal state of affairs, the question you may be interested in investigating is: How can you personally increase your creative ability?

11.4.4.6. Improving Your Creative Ability. In the light of our theory, the apparent answer to this question is: get rid of your inhibitions–those insecurities that block your creativity. Unfortunately, in opposition to many psychological theories, I don't believe this can be done.

An adult with his inhibitions is like a ruler that was misused in its early life and stretched out of proportion by 10 percent. You can put the ruler on the couch, listen to it sympathetically, have it relive its traumatic experiences, and so on, but it will continue to say that there are 13.2 inches in every foot.

However, all isn't lost. Once it's recognized of what the ruler's distortion consists, it's possible to compensate for the distortion and come up with a true measurement. Similarly, once you know what your inhibitions are, you can compensate for them. I don't say that during the five percent of the time when you're under pressure, you're not going to react as you're programmed, because you are. I am saying that, during the other 95 percent of your life, when you can take a more measured approach to things, compensating for your inhibitions will make you more creative.

So how do you find out what you inhibitions are? A book that you should find helpful in this area is Edward deBono's *The Five-Day Course in Thinking* [2]. This book consists of a well-designed series of exercises, whose purpose isn't so much to reach a solution as to allow you to study your own habits of thought as you go through the problem-solving process. You thus learn your strengths and your weaknesses, and specifically, you become familiar with your own personal and unique blocks against creativity. As you learn about yourself, the exercises give you the opportunity to compensate for your inhibitions and thus to become more creative. No one who makes serious use of this book can fail to benefit from it.

11.4.4.7. Creative Procedures. There are some recognized techniques for trying to get insight. We'll review these techniques here.

One technique is fluency—the generation of a large number of ideas. Don't immediately try to solve the problem. First, try to come up with as many ideas about the problem and the information you've collected as possible.

One way of generating ideas is brainstorming. Although brainstorming is generally done by a group of people, you can also do it by yourself. Just remember the basic rule: no criticism of any ideas while in the brainstorming mode.

Our description of brainstorming brings out two other techniques of creative thinking: elaboration and transformation. Elaboration is the embellishment of a basic idea with details. Transformation is the combination of things in new ways. Both can be done methodically. After or during a brainstorming session, you can deliberately try to embellish every idea generated with as much detail as possible. You can also make a matrix of your ideas and methodically investigate each possible combination of ideas.

Another technique for generating a large number of ideas is extended effort. In this regard, here's some advice from deBono.

> Every decision is made with some degree of uncertainty. Confidence in a decision does not depend on the lack of any alternative, for that might indicate a lack of imagination, but on the ability to see many alternatives, all of which can be rejected [3, p. 149].

A fourth creative thinking technique is flexibility—changing your frame of reference. Look at things from as many points of view as possible. The merchant's daughter used the technique of flexibility when she changed her frame of reference from the color of the pebble she had to select to the color of the pebble left in the bag.

A fifth technique is to try to develop general principles. It's probably better to use an incorrect principle and modify it as required than to avoid principles altogether. For example, the general principle that can be derived from the example of the merchant's daughter might be stated as follows.

> In a situation in which chance is supposed to hold sway, but where in fact the game has been rigged, it's often possible to turn this fact to your advantage.

Once the general principle has been stated, other situations to which it applies often come to mind. For example, gamblers say that the way to beat a crooked wheel is to make small bets against the big bettor. The theory is that when the house exercises its influence over the wheel, it will do so to defeat the big bettor, not the little ones.

Ultimately, all generalizations are based on experience. Past experience is most useful in solving new problems if the attempt to generalize on the experience is made at the time the experience occurs. So part of creative thinking is discipline. When you've solved a problem, or when you've seen someone solve a problem, think about it from the point of view of what general principles were applied. Where else have you seen these same general principles applied successfully? What common characteristics do problems in which the principles apply have? In this way you organize your experience and prepare yourself for the problems you haven't yet faced.

In general, the more your past experience is related to the problem at hand, the easier the problem is to solve. Thus, after you've designed a number of programs, designing the next one isn't much of a challenge. It's that first program design that's formidable. So here's some advice: If the problem you're facing is unfamiliar to you, seek out the advice of people who have contended with similar problems.

If prior similar experiences aren't available, analogies are sometimes helpful. For example, consider the following problem.

> A harmful tumor inside the body of a patient can be eradicated by a sufficient concentration of X-rays. However, a

> beam of the required strength would also destroy all intervening tissue. On the other hand, a beam weak enough not to harm the surrounding tissue would be too weak to destroy the tumor. *What method should be used to destroy the tumor?*

Here's a hint: Think of how core storage operates.

To destroy the tumor, several weak beams proceeding from different points of origin are made to intersect at the tumor's location. Anyone familiar with the operation of core storage should have an advantage in solving this problem, since core storage operates by means of weak currents combining to operate on cores at the intersections of the currents.

Finally, the makeup of a good solution is something we can never predict. We must rely on the natural creative ability of our minds to come up with this combination. So vary periods of concentration on the problem with periods of, if you will, benevolent neglect. The common experience of going to bed with a problem and waking up with the solution emphasizes the importance of this technique. However, don't think you can solve problems by ignoring them. Benevolent neglect works only after extended periods of deep immersion in the problem. Then your subconscious has the material it needs to work on while your conscious attention is elsewhere.

11.4.5. Testing the Ideas

We must always remember that, to be creative, an idea must be practical; it must work; it must, in fact, solve the problem. So we must subject every idea to the test, because until we do, we don't know whether it's a great idea or just a flight of fancy. The great contribution of scientific method is recognition that it's thinking *plus* testing which creates knowledge.

Ideas are tested both during their development and after a tentative solution has been selected. During idea development, typical idea testing consists of the use of deduction to eliminate unsuitable ideas. However, a tentative solution is eventually selected, and this solution must be tested, to see if it holds up, before you can conclude that your problem is solved.

11.5. APPLICATION TO DATA PROCESSING

Creative thinking is used at all points in the definition, design, and construction of a data processing system. As an example, consider the design of any data processing system.

To determine the best design, you must have an objective against which to measure alternatives. Determination of the objective of a system is equivalent to problem definition. The more precise the definition, the greater the likelihood of a successful design.

When doing the design, you should develop alternatives and ultimately fix on the design that incorporates the best features of the alternatives. Remember: Don't concentrate on solving the problem right away. First generate ideas. Too often poor design is a function of adopting the first approach considered. And don't forget to get the advice and consultation of other design experts.

The good system designers with which I've had the good fortune to be acquainted use the same technique in coming up with a good design. First they specify the most grubby, quick and dirty, down-to-earth, pound-at-it-until-it's-done solution to the problem that they can think of. Next they outline the most blue-sky, beyond-the-state-of-the-art, exotic solution they can conceive of. This sets the limits of the possibilities. Finally, they sketch out various alternative solutions that seem to represent significant gradations between these extremes. (This is an application of the principle of fluency.) After identifying these alternatives, they start filling in details and rearranging parts of alternatives to come up with new configurations. (This is the application of the principles of elaboration and transformation.) As this process continues, certain alternatives are demonstrated to be undesirable, while others reveal various combinations of desirable and undesirable characteristics. Ultimately, either an optimum design emerges, or the field is narrowed to a few competing contenders, at which point the search for solutions ends and the pursuit of goal refinement begins, so a principle for choosing between contenders can be established.

You don't know whether your design works until the system is put into production. Even then, you'll never get a clear test of the effectiveness of your design efforts unless, as part of that effort, you include, in the system, features that will give you feedback on how well your design is meeting its objectives in practice. Incidentally, to build in these feedback mechanisms, it's once more necessary to have clearly defined objectives so you have something against which to measure the effectiveness of your design.

BIBLIOGRAPHY

1. Black, Max. *Critical Thinking.* Englewood Cliffs, N.J.: Prentice-Hall, 1946. From this book we have taken the ways of identifying authorities, the problem concerning light waves from Sirius, and the problem of destroying a tumor with X rays.

2. deBono, Edward. *The Five-Day Course in Thinking.* New York: Basic Books, 1967.

3. deBono, Edward. *New Think.* New York: Basic Books, 1968.
From this book we have taken the anecdote concerning the merchant, his daughter, and the money lender; and the problem concerning the bomber hydraulic system.

4. Hare, Van Court, Jr. *Systems Analysis: A Diagnostic Approach.* New York: Harcourt Brace Jovanovich, 1967.
From this book we have taken the problem concerning the five houses and their occupants.

5. Kepner, Charles H., and Tregoe, Benjamin B. *The Rational Manager.* New York: McGraw-Hill, 1965.

6. Osborne, Alex F. *Applied Imagination.* New York: Charles Scribner's Sons, 1963.
The concept of brainstorming comes from Osborne.

7. Parnes, Sidney J. *Creative Behavior Guidebook.* New York: Charles Scribner's Sons, 1967.
The concept of extended effort comes from Parnes.

8. Steiner, Gary A. *The Creative Organization.* Chicago: The University of Chicago Press, 1965.
The list of traits of creative people was compiled from Steiner and Parnes.

9. Upton, Albert, and Samson, Richard W. *Creative Analysis.* New York: Dutton, 1961.
The problem concerning the five men and their marital status is taken from Upton and Samson.

EXERCISES

1. Joyce is three years older than Debbie, and their mother, Connie, is three years older than the daughters' combined ages. Their father, Bill, is five years older than their mother. Another daughter, Patti, is twice the age of Debbie, and four years older than Anne, another daughter. Debbie owns a rag doll that is two years younger than Debbie. All the ages combined, except the rag doll's, total 98. How old is each member of the family?

2. A father wished to leave his fortune to the most intelligent of his three sons. He said to them: "I shall presently take each one of you away separately and paint either a white or a blue mark on each of your foreheads; and none of you will have any chance to know the color of the mark on his own head. Then I shall bring you together again, and anybody who is able to see two blue marks on the heads of his companions is to laugh. The first of you to *deduce* his own color is to raise his hand, and on convincing me that his solution is correct will become my heir." After all three had agreed to the conditions, the father took them apart and painted a *white* mark on each forehead. When they met again, there was silence for some time, at the end of which

the youngest brother raised his hand, saying, "I'm white." How was he able to deduce the color of the mark on his forehead? (From Max Black, *Critical Thinking,* © 1946. Reprinted by permission of Prentice-Hall, Inc., Englewood Cliffs, New Jersey).

3. Comment on the following argument: Usually when you find a traffic jam you also find a policeman directing traffic. Therefore most traffic jams are caused by policemen.

12

Cost Benefit Analysis

A cost benefit analysis is a procedure for collecting and organizing all the facts pertinent to determining the economic feasibility of developing a proposed data processing system. It's a major part of the system proposal, and as system development progresses through the analysis and design, this original cost benefit analysis is often revised. In this chapter we look into the nature of a cost benefit analysis and the method of developing one.

12.1. FEASIBILITY

As soon as the development of a data processing system is contemplated, the question of the *feasibility* of the development is raised. As first stated in the McKinsey report, *Unlocking the Computer's Profit Potential,* there are three parts to the question of feasibility.

1. Is developing the system possible? (Is it technically feasible?)
2. If we develop the system:
   a. Will it be used? (Is it operationally feasible?)
   b. Will it pay off? (Is it economically feasible?)

12.1.1. Technical Feasibility

The question of technical feasibility is concerned with whether the know-how to develop the system exists. At one time there was a real question as to whether a system being contemplated was technically feasible. For example, before the development of direct-access, mass storage devices, realtime systems weren't technically feasible. But with today's state of the art, the development of most conceivable data processing systems is technically feasible. The hardware and the know-how to use the hardware exist. It may be prohibitively expensive to acquire the hardware and to hire the know-how (if it doesn't exist in your firm) to put the hardware to work, but this is a question of economic rather than technical feasibility.

12.1.2. Operational Feasibility

The question of operational feasibility is a real one. The story of the terminal that sits unused on the executive's credenza is common, and the number of unopened computer-generated reports is legion. The expense of an unused system is never zero, and the benefit of an unused system is always zero. Thus a system that isn't operationally feasible is never economically feasible, regardless of the weight of the cost-justification studies supporting its development.

Determination of operational feasibility is crucial to the credibility of any demonstration of economic feasibility. The issue of operational feasibility is frequently not given adequate consideration, and its importance can't be overemphasized. Nevertheless, it's not the subject of this chapter. In the remainder of this chapter we assume the technical and operational feasibility of the proposed system.

12.1.3. Economic Feasibility

Until the millennium, when there will be enough of everything for everyone, we're going to be faced with the problem of allocating scarce resources among competing needs. The method of making this allocation is the fundamental economic problem, and this is the question we face when we ask, "If we develop this system, will it pay off?"

However, the economic question is more complicated than the simple determination of whether a system will pay off. Some investments have better payoffs than others. Firms have limited resources, and they want to put their money where they get the biggest return. Development of a particular data processing system may be economically feasible, but if it's competing with a new product, an expansion of marketing activity, or even some other system where the payoff is bigger, it makes good economic sense to go for the bigger payoff. So we want not only to determine whether a system is economically feasible but also to develop an indicator of the profitability of the system on a common yardstick, so we can compare the attractiveness of developing the system with other competing investment opportunities.

When considering the development of a proposed data processing system, the question of economic feasibility is one we can never responsibly evade. It's to the method of answering this question that this chapter is addressed.

As we've already implied, investment in a data processing system is no different from any other capital investment. Consequently, to investigate the economic feasibility of data processing systems, we have to become familiar with some financial concepts.

12.2. FIXED AND VARIABLE COSTS

The first financial concept we'll look at is that of fixed and variable costs.

Companies commonly divide their costs into two categories: *fixed costs,* which don't vary with changes in volume of activity, and *variable costs,* which do. For example, interest on bonds and property taxes are fixed costs, because they go on regardless of production level. On the other hand, cost of materials and wages of workers on the production line are variable costs, because they vary with the production level.

The concept of fixed cost is related to periods of time. With the exception of the cost of the residual management always required by a firm, all costs are variable in the long run. However, equipment, such as plants and machines, does have the characteristic of a fixed cost during its life; the commitment to the facilities has been made and is difficult to avoid.

12.3. DIFFERENTIAL COSTS

When weighing the pros and cons of an investment, both the cost of the investment and the benefits to be derived from the investment must be considered. The costs taken into consideration are the *differential costs*—costs that wouldn't be present if the investment wasn't made. Usually, the differential costs of an investment are the variable costs related to the investment.

12.4. ECONOMIC LIFE

To determine the costs and benefits of an investment, it's necessary to fix the economic life of the investment. The *economic life* of an investment is the number of years the investment can be expected to yield benefits.

For some investments, the economic life is easy to determine. For example, the number of years a machine can be used before it wears out can be fixed quite precisely in some cases. In other cases, the economic life of an investment may be indefinitely long. Nevertheless, even in such cases there's some upper limit beyond which it's impractical to project economic life.

An investment's economic life is always a projection into the future. With each additional year of projection, the reliability of the projection becomes more uncertain. Changes in technology or market may wipe out future anticipated benefits. The relative stability of the business in which the investment is made has an influence on the maximum number of years for which it's practical to assume an economic life.

In any case, once the economic life of an investment is fixed, it becomes possible to classify the costs involved as fixed or variable. The costs of existing facilities that will be present during the economic life of the investment are fixed. The other costs are variable.

12.5. SUNK COSTS

No fixed cost related to an investment is considered in determining the investment's merits. Such costs are called *sunk costs,* because they have been "sunk" into the facilities prior to the investment opportunity. For example, if you're contemplating the use of a presently idle factory that you own in producing a new product, the cost of originally acquiring the factory isn't taken into consideration in determining whether to produce the product. You've already sunk your money into the factory, and this prior investment has no bearing on whether production of the new product is a good investment.

12.6. DEVELOPMENT AND OPERATION OF A DATA PROCESSING SYSTEM AS AN INVESTMENT

When considering the development of a data processing system, it's not uncommon to run up against a judgment phrased somewhat as follows.

> Of course development of this system is economically feasible. We're going to get benefits from it. It's going to be run on equipment we already have in the computer center by people we already have working there. And it's going to be developed by people we already have working in the system development department.

Analysis of such a comment indicates that the speaker considers the cost of the equipment and the salaries of computer center and system development people to be sunk costs, which therefore shouldn't be considered in deciding whether development of this system will pay off. Consequently system costs are negligible, and since the system will yield benefits, it's apparent on the face of it that developing it will pay off.

This approach is wrongheaded on both counts.

1. The equipment used in a computer center is highly modular. There's no need to have more than can be justified. Moreover, there's no requirement to make a long-term commitment to the equipment. It can be rented under an agreement allowing for cancellation on short notice. Even if the equipment was originally purchased, there's an active second-hand market on which attractive returns on sales can be obtained.
2. Humanitarian considerations aside, from an economic point of view, people are the epitome of a variable cost. As companies in serious financial binds quickly discover, people can be disposed of by simply saying the word.

Development of a data processing system is an investment question in which all costs are variable and must be considered. As a consequence, development of the system must compete with all the other opportunities a firm has in which to invest its scarce capital resources. If a firm can make more money by developing a new product, by expanding its marketing strategy, or by adding to its production capacity than it can by developing a particular data processing system, and if the firm has insufficient funds to undertake all these activities, then the rational choice is to leave the system undeveloped. The purpose of this chapter is to determine the basis on which this choice is made.

12.7. THE INVESTMENT SITUATION

The typical investment situation can be described graphically as shown in Figure 12.1. First there's a relatively steep negative flow of cash as the development involved in the investment is made. This

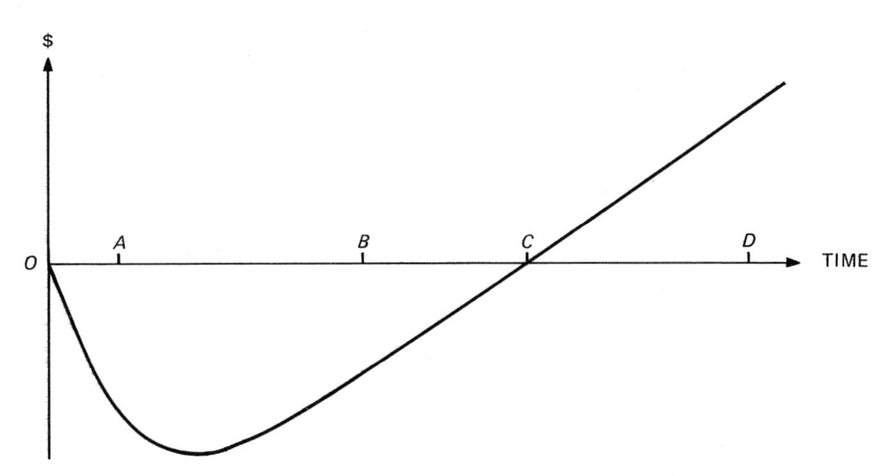

Figure 12.1 The investment situation.

might represent the construction of production facilities. Then as the development effort nears completion, the money spent on development tapers off and operation of the facilities developed begins. There are operating expenses, but operation also returns benefits. In the case of a newly constructed factory, the benefits would be represented by the revenue obtained through the sale of factory output. In a sound investment, benefits exceed operating expenses, and over a period of time the cash flow curve bottoms out, turns around, and begins a slower but steady climb that first wipes out the indebtedness created by the development expense and then begins to generate a positive return on the investment.

One aspect of the attractiveness of an investment inheres in how the cash flow curve in Figure 12.1 fits into the investment's economic life. If the economic life runs from the origin, O, on the time line to point A, the investment is an unmitigated disaster. Most investments wouldn't have such a total projected economic life, but all good investments project a point where cash flow is supposed to turn around. If this point is passed with no diminution of the rate of cash outflow for development expense, there's no telling to what level the cash flow may plunge.

If the investment's economic life extends from O to point B, the negative cash flow has been stemmed, but there's insufficient time to take advantage of the benefits that exceed operating expense. As a consequence, the investment never pays off.

If the economic life extends from O to point C, enough benefits have been realized to balance the initial investment, and the investment breaks even. Understandably, this point is called the *breakeven point.* Only when the economic life extends beyond the breakeven point to, for example, point D does the investment yield a return.

12.8. CASH FLOW

Figure 12.1 is a graphic picture of a cash flow analysis. In table form (the more standard way of doing a cash flow analysis) the picture might look as is shown in Figure 12.2.

Roughly speaking, a cash flow analysis, such as the one shown in Figure 12.2, will always demonstrate whether an investment pays off. It's a matter of whether the ending balance for the economic life of the investment is positive or negative.

12.9. RANKING INVESTMENTS

The question that must be posed for a cash flow analysis with a positive ending balance is, How well does the investment pay off? Only when we can answer this question for a number of investments is it possible to rank competing investments in terms of relative desirability. Three ways of approaching this question are in terms of:

1. Payback period
2. Return on investment
3. Present value

12.9.1. Payback Period

The *payback period* is the period of time required for the investment to break even. In Figure 12.1 the payback period is represented by the time line extending from the origin O to the breakeven point, C. In Figure 12.2 the payback period is somewhat more than seven quarters. In general, if an investment involves development costs of D and if annual revenues exceed annual operating costs by P, then the payback period, T, can be found by using the formula

$$T = \frac{D}{P}$$

The shorter the payback period, the better the investment. For example, consider the following two investments.

Investment	A	B
Development costs	\$40,000	\$50,000
Annual benefit	20,000	30,000

In the case of investment A, the payback period is:

$$\frac{40,000}{20,000} = 2 \text{ years}$$

QUARTER	1	2	3	4	5	6	7	8
BEGINNING BALANCE	—	(56)	(108)	(136)	(112)	(83)	(50)	(10)
RECEIPTS	—	—	—	54	61	67	74	74
DEVELOPMENT COSTS	56	40	—	—	—	—	—	—
OPERATING COSTS	—	12	28	30	32	34	34	30
ENDING BALANCE	(56)	(108)	(136)	(112)	(83)	(50)	(10)	34

QUARTER	9	10	11	12	13	14	15	16
BEGINNING BALANCE	34	65	96	133	149	165	187	227
RECEIPTS	61	61	61	40	40	40	40	20
DEVELOPMENT COSTS	—	—	—	—	—	—	—	—
OPERATING COSTS	30	30	24	24	24	18	—	—
ENDING BALANCE	65	96	133	149	165	187	227	247

Figure 12.2 Cash flow analysis in thousands of dollars for an investment with a four-year economic life.

In the case of investment B, the payback period is:

$$\frac{50{,}000}{30{,}000} = 1.67 \text{ years}$$

Therefore, in terms of payback period, B is the better investment.

12.9.2. Return on Investment

The return on investment is an expression of the benefits returned by an investment in terms of a rate per year.

For example, in Figure 12.2, total benefits are $247,000 received over a four-year period. Thus average annual benefits are $61,750.

Total development costs, which are $96,000, represent the investment made. The general approach here is to regard this investment as being repaid over the economic life of the investment in a straight-line fashion, so that at the beginning of the economic life the total investment is outstanding, while at the end of economic life the investment is totally repaid. (See Figure 12.3.) Under such conditions, the average investment outstanding is one half the sum of the investment at the beginning of economic life and the investment at the end of economic life. Since the investment at the beginning of economic life is the total investment and the investment at the end of economic life is zero, average investment outstanding boils down to one half the total investment. In the case of Figure 12.2, the average investment is thus one half of $96,000, or $48,000.

The return on investment is average annual benefits as a percentage of average annual investment.

$$\frac{\$61{,}750}{\$48{,}000} = 129\%$$

A very good return on investment.

In general, if an investment involves development costs of D, if annual revenues exceed annual operating costs by P, and if economic life is N years, then the return on investment, R, can be found by using the formula

$$R = \frac{P - \frac{D}{N}}{\frac{D}{2}} = \frac{2\left(P - \frac{D}{N}\right)}{D}$$

The higher the return on investment, the better the investment. For example, consider the following two investments.

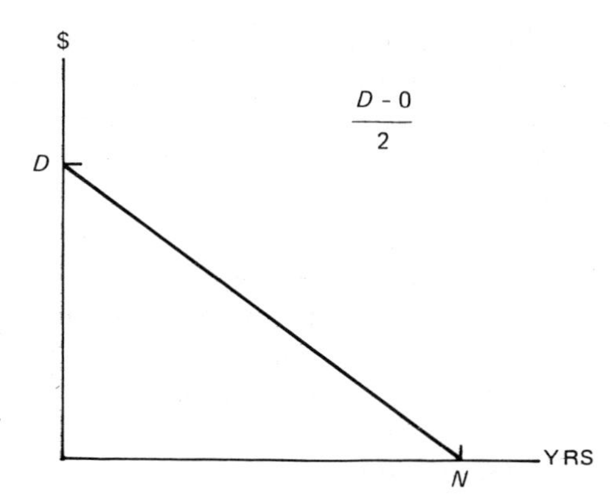

Figure 12.3 Straight-line depreciation of development costs.

Investment	A	B
Development costs	\$40,000	\$50,000
Annual benefit	\$20,000	\$30,000
Economic life	5 years	3 years

In the case of investment A, the return on investment is:

$$\frac{2\left(20{,}000 - \dfrac{40{,}000}{5}\right)}{40{,}000} = 60\%$$

In the case of investment B, the return on investment is:

$$\frac{2\left(30{,}000 - \dfrac{50{,}000}{3}\right)}{50{,}000} = 53.3\%$$

Therefore, in terms of return on investment, A is the better investment.

12.9.3. Comparison of Return on Investment and Payback Period

By now you may be confused. We have shown that in terms of return on investment, investment A is the better of our two example investments, while in terms of payback period, investment B is to be preferred. How do we resolve this contradiction?

As we've already determined, organizations are concerned with investing their limited capital resources where they can make the most money. From this point of view, return on investment provides pertinent guidance. In the long run, an investment with a longer payback period may be a better money maker than one with a shorter payback period.

On the other hand, concern with the payback period isn't inappropriate. It focuses on when the investment is going to start making money, a consideration ignored when directing attention to return on investment. And this concern with timing is relevant, since a dollar today is worth more than a dollar next year, even when inflation isn't a factor.

What we need is a technique for evaluating investments that takes into consideration both how much money an investment is going to generate and when this money is going to be generated. The present value of benefits is such a technique.

12.9.4. Present Value

In these inflationary times, we all recognize that a dollar today is worth more than a dollar next year. But even if inflation isn't a factor, a dollar today is still worth more than a dollar next year.

12.9.4.1. The Time Value of Money. Suppose you deposit $1 in a savings account that pays 8 percent interest. At the end of a year, your account will contain $1.08. Thus you can see that after one year your dollar will be worth $1.08.

But look at the situation from the other side. The fact is that $1.08 one year from now is worth only $1 today. Thus money one year from now is worth less than the same amount of money today. This is the *time value* of money. Money is an asset we expect to increase in value over time.

In our example, $1 is the *present value* of $1.08 one year from now. In general, if i is the annual rate of return at which we expect our money to increase, then the present value of $1 to be received n years from now is:

$$\frac{1}{(1 + i)^n}$$

This isn't an easy formula to manipulate, so rather than use it directly, we use a table of values generated by the formula. One such table is shown in Figure 12.4. Using this table we can determine, for example, that at 8 percent, $1 one year from now is presently worth somewhat less than $0.93.

An inspection of Figure 12.4 shows another reason why there's not much point in considering too long an economic life when evaluating an investment. This reason is that not very many years have to pass before the present value of a future benefit becomes too small to be considered significant. For example, at 15 percent, the present

Years Hence	1%	2%	4%	6%	8%	10%	12%	14%	15%	16%	18%	20%	22%	24%	25%	26%	28%	30%	35%	40%	45%	50%
1	0.990	0.980	0.962	0.943	0.926	0.909	0.893	0.877	0.870	0.862	0.847	0.833	0.820	0.806	0.800	0.794	0.781	0.769	0.741	0.714	0.690	0.667
2	0.980	0.961	0.925	0.890	0.857	0.826	0.797	0.769	0.756	0.743	0.718	0.694	0.672	0.650	0.640	0.630	0.610	0.592	0.549	0.510	0.476	0.444
3	0.971	0.942	0.889	0.840	0.794	0.751	0.712	0.675	0.658	0.641	0.609	0.579	0.551	0.524	0.512	0.500	0.477	0.455	0.406	0.364	0.328	0.296
4	0.961	0.924	0.855	0.792	0.735	0.683	0.636	0.592	0.572	0.552	0.516	0.482	0.451	0.423	0.410	0.397	0.373	0.350	0.301	0.260	0.226	0.198
5	0.951	0.906	0.822	0.747	0.681	0.621	0.567	0.519	0.497	0.476	0.437	0.402	0.370	0.341	0.328	0.315	0.291	0.269	0.223	0.186	0.156	0.132
6	0.942	0.888	0.790	0.705	0.630	0.564	0.507	0.456	0.432	0.410	0.370	0.335	0.303	0.275	0.262	0.250	0.227	0.207	0.165	0.133	0.108	0.088
7	0.933	0.871	0.760	0.665	0.583	0.513	0.452	0.400	0.376	0.354	0.314	0.279	0.249	0.222	0.210	0.198	0.178	0.159	0.122	0.095	0.074	0.059
8	0.923	0.853	0.731	0.627	0.540	0.467	0.404	0.351	0.327	0.305	0.266	0.233	0.204	0.179	0.168	0.157	0.139	0.123	0.091	0.068	0.051	0.039
9	0.914	0.837	0.703	0.592	0.500	0.424	0.361	0.308	0.284	0.263	0.225	0.194	0.167	0.144	0.134	0.125	0.108	0.094	0.067	0.048	0.035	0.026
10	0.905	0.820	0.676	0.558	0.463	0.386	0.322	0.270	0.247	0.227	0.191	0.162	0.137	0.116	0.107	0.099	0.085	0.073	0.050	0.035	0.024	0.017
11	0.896	0.804	0.650	0.527	0.429	0.350	0.287	0.237	0.215	0.195	0.162	0.135	0.112	0.094	0.086	0.079	0.066	0.056	0.037	0.025	0.017	0.012
12	0.887	0.788	0.625	0.497	0.397	0.319	0.257	0.208	0.187	0.168	0.137	0.112	0.092	0.076	0.069	0.062	0.052	0.043	0.027	0.018	0.012	0.008
13	0.879	0.773	0.601	0.469	0.368	0.290	0.229	0.182	0.163	0.145	0.116	0.093	0.075	0.061	0.055	0.050	0.040	0.033	0.020	0.013	0.008	0.005
14	0.870	0.758	0.577	0.442	0.340	0.263	0.205	0.160	0.141	0.125	0.099	0.078	0.062	0.049	0.044	0.039	0.032	0.025	0.015	0.009	0.006	0.003
15	0.861	0.743	0.555	0.417	0.315	0.239	0.183	0.140	0.123	0.108	0.084	0.065	0.051	0.040	0.035	0.031	0.025	0.020	0.011	0.006	0.004	0.002
16	0.853	0.728	0.534	0.394	0.292	0.218	0.163	0.123	0.107	0.093	0.071	0.054	0.042	0.032	0.028	0.025	0.019	0.015	0.008	0.005	0.003	0.002
17	0.844	0.714	0.513	0.371	0.270	0.198	0.146	0.108	0.093	0.080	0.060	0.045	0.034	0.026	0.023	0.020	0.015	0.012	0.006	0.003	0.002	0.001
18	0.836	0.700	0.494	0.350	0.250	0.180	0.130	0.095	0.081	0.069	0.051	0.038	0.028	0.021	0.018	0.016	0.012	0.009	0.005	0.002	0.001	0.001
19	0.828	0.686	0.475	0.331	0.232	0.164	0.116	0.083	0.070	0.060	0.043	0.031	0.023	0.017	0.014	0.012	0.009	0.007	0.003	0.002	0.001	
20	0.820	0.673	0.456	0.312	0.215	0.149	0.104	0.073	0.061	0.051	0.037	0.026	0.019	0.014	0.012	0.010	0.007	0.005	0.002	0.001	0.001	
21	0.811	0.660	0.439	0.294	0.199	0.135	0.093	0.064	0.053	0.044	0.031	0.022	0.015	0.011	0.009	0.008	0.006	0.004	0.002	0.001		
22	0.803	0.647	0.422	0.278	0.184	0.123	0.083	0.056	0.046	0.038	0.026	0.018	0.013	0.009	0.007	0.006	0.004	0.003	0.001	0.001		
23	0.795	0.634	0.406	0.262	0.170	0.112	0.074	0.049	0.040	0.033	0.022	0.015	0.010	0.007	0.006	0.005	0.003	0.002	0.001			
24	0.788	0.622	0.390	0.247	0.158	0.102	0.066	0.043	0.035	0.028	0.019	0.013	0.008	0.006	0.005	0.004	0.003	0.002	0.001			
25	0.780	0.610	0.375	0.233	0.146	0.092	0.059	0.038	0.030	0.024	0.016	0.010	0.007	0.005	0.004	0.003	0.002	0.001	0.001			
26	0.772	0.598	0.361	0.220	0.135	0.084	0.053	0.033	0.026	0.021	0.014	0.009	0.006	0.004	0.003	0.002	0.002	0.001				
27	0.764	0.586	0.347	0.207	0.125	0.076	0.047	0.029	0.023	0.018	0.011	0.007	0.005	0.003	0.002	0.002	0.001	0.001				
28	0.757	0.574	0.333	0.196	0.116	0.069	0.042	0.026	0.020	0.016	0.010	0.006	0.004	0.002	0.002	0.002	0.001	0.001				
29	0.749	0.563	0.321	0.185	0.107	0.063	0.037	0.022	0.017	0.014	0.008	0.005	0.003	0.002	0.002	0.001	0.001	0.001				
30	0.742	0.552	0.308	0.174	0.099	0.057	0.033	0.020	0.015	0.012	0.007	0.004	0.003	0.002	0.001	0.001	0.001					
40	0.672	0.453	0.208	0.097	0.046	0.022	0.011	0.005	0.004	0.003	0.001	0.001										
50	0.608	0.372	0.141	0.054	0.021	0.009	0.003	0.001	0.001	0.001												

Figure 12.4 Present value of $1.

value of $1 received six years from now is only a fraction more than $0.43.

Now let's consider investment A with the following characteristics.

Annual benefit	$20,000
Economic life	5 years
Rate of return	15 percent

Here we're going to receive $20,000 one year from now, another $20,000 two years from now, a third $20,000 three years from now, a fourth $20,000 four years from now, and a fifth $20,000 five years from now. Figure 12.4 tells us that $1 received one year from now has a present value of $0.870; $1 received two years from now, a present value of $0.756; $1 received three years from now, a present value of $0.658; $1 received four years from now, a present value of $0.572; and $1 received five years from now, a present value of $0.497. Thus the $20,000 received one year from now has a present value of $17,400 (0.870 X $20,000); the $20,000 received two years from now, a present value of $15,120; the $20,000 received three years from now, a present value of $13,160; the $20,000 received four years from now, a present value of $11,440; and the $20,000 received five years from now, a present value of $9,940. Summing these individual present values tells us that the benefits of investment A have a present value of:

$17,400 + 15,120 + 13,160 + 11,440 + 9,940 = $67,060

Many investments have the characteristic of yielding a steady return year after year. Investment A has this characteristic, in this case the return being $20,000 a year. Since this situation is common, the work needed to calculate the present value of the benefits of such an investment can be shortened by using the table in Figure 12.5. Each entry in this table is just the sum of the entries for the given year and all preceding years from Figure 12.4. Thus, the entry for five years at 15 percent is just the sum of 0.870, 0.756, 0.658, 0.572, and 0.497. Using Figure 12.5, we can easily calculate the present value of the benefits of investment A as:

3.352 ($20,000) = $67,060

Many investments, such as product lines and data processing systems, don't yield benefits at the end of each year of economic life. Instead, the benefit is generated continuously over economic life. There are present value tables constructed on the assumption of continuous return. However, this is a refinement that typically isn't resorted to.

Years (N)	1%	2%	4%	6%	8%	10%	12%	14%	15%	16%	18%	20%	22%	24%	25%	26%	28%	30%	35%	40%	45%	50%
1.......	0.990	0.980	0.962	0.943	0.926	0.909	0.893	0.877	0.870	0.862	0.847	0.833	0.820	0.806	0.800	0.794	0.781	0.769	0.741	0.714	0.690	0.667
2.......	1.970	1.942	1.886	1.833	1.783	1.736	1.690	1.647	1.626	1.605	1.566	1.528	1.492	1.457	1.440	1.424	1.392	1.361	1.289	1.224	1.165	1.111
3.......	2.941	2.884	2.775	2.673	2.577	2.487	2.402	2.322	2.283	2.246	2.174	2.106	2.042	1.981	1.952	1.923	1.868	1.816	1.696	1.589	1.493	1.407
4.......	3.902	3.808	3.630	3.465	3.312	3.170	3.037	2.914	2.855	2.798	2.690	2.589	2.494	2.404	2.362	2.320	2.241	2.166	1.997	1.849	1.720	1.605
5.......	4.853	4.713	4.452	4.212	3.993	3.791	3.605	3.433	3.352	3.274	3.127	2.991	2.864	2.745	2.689	2.635	2.532	2.436	2.220	2.035	1.876	1.737
6.......	5.795	5.601	5.242	4.917	4.623	4.355	4.111	3.889	3.784	3.685	3.498	3.326	3.167	3.020	2.951	2.885	2.759	2.643	2.385	2.168	1.983	1.824
7.......	6.728	6.472	6.002	5.582	5.206	4.868	4.564	4.288	4.160	4.039	3.812	3.605	3.416	3.242	3.161	3.083	2.937	2.802	2.508	2.263	2.057	1.883
8.......	7.652	7.325	6.733	6.210	5.747	5.335	4.968	4.639	4.487	4.344	4.078	3.837	3.619	3.421	3.329	3.241	3.076	2.925	2.598	2.331	2.108	1.922
9.......	8.566	8.162	7.435	6.802	6.247	5.759	5.328	4.946	4.772	4.607	4.303	4.031	3.786	3.566	3.463	3.366	3.184	3.019	2.665	2.379	2.144	1.948
10.......	9.471	8.983	8.111	7.360	6.710	6.145	5.650	5.216	5.019	4.833	4.494	4.192	3.923	3.682	3.571	3.465	3.269	3.092	2.715	2.414	2.168	1.965
11.......	10.368	9.787	8.760	7.887	7.139	6.495	5.937	5.453	5.234	5.029	4.656	4.327	4.035	3.776	3.656	3.544	3.335	3.147	2.752	2.438	2.185	1.977
12.......	11.255	10.575	9.385	8.384	7.536	6.814	6.194	5.660	5.421	5.197	4.793	4.439	4.127	3.851	3.725	3.606	3.387	3.190	2.779	2.456	2.196	1.985
13.......	12.134	11.343	9.986	8.853	7.904	7.103	6.424	5.842	5.583	5.342	4.910	4.533	4.203	3.912	3.780	3.656	3.427	3.223	2.799	2.468	2.204	1.990
14.......	13.004	12.106	10.563	9.295	8.244	7.367	6.628	6.002	5.724	5.468	5.008	4.611	4.265	3.962	3.824	3.695	3.459	3.249	2.814	2.477	2.210	1.993
15.......	13.865	12.849	11.118	9.712	8.559	7.606	6.811	6.142	5.847	5.575	5.092	4.675	4.315	4.001	3.859	3.726	3.483	3.268	2.825	2.484	2.214	1.995
16.......	14.718	13.578	11.652	10.106	8.851	7.824	6.974	6.265	5.954	5.669	5.162	4.730	4.357	4.033	3.887	3.751	3.503	3.283	2.834	2.489	2.216	1.997
17.......	15.562	14.292	12.166	10.477	9.122	8.022	7.120	6.373	6.047	5.749	5.222	4.775	4.391	4.059	3.910	3.771	3.518	3.295	2.840	2.492	2.218	1.998
18.......	16.398	14.992	12.659	10.828	9.372	8.201	7.250	6.467	6.128	5.818	5.273	4.812	4.419	4.080	3.928	3.786	3.529	3.304	2.844	2.494	2.219	1.999
19.......	17.226	15.678	13.134	11.158	9.604	8.365	7.366	6.550	6.198	5.877	5.316	4.844	4.442	4.097	3.942	3.799	3.539	3.311	2.848	2.496	2.220	1.999
20.......	18.046	16.351	13.590	11.470	9.818	8.514	7.469	6.623	6.259	5.929	5.353	4.870	4.460	4.110	3.954	3.808	3.546	3.316	2.850	2.497	2.221	1.999
21.......	18.857	17.011	14.029	11.764	10.017	8.649	7.562	6.687	6.312	5.973	5.384	4.891	4.476	4.121	3.963	3.816	3.551	3.320	2.852	2.498	2.221	2.000
22.......	19.660	17.658	14.451	12.042	10.201	8.772	7.645	6.743	6.359	6.011	5.410	4.909	4.488	4.130	3.970	3.822	3.556	3.323	2.853	2.498	2.222	2.000
23.......	20.456	18.292	14.857	12.303	10.371	8.883	7.718	6.792	6.399	6.044	5.432	4.925	4.499	4.137	3.976	3.827	3.559	3.325	2.854	2.499	2.222	2.000
24.......	21.243	18.914	15.247	12.550	10.529	8.985	7.784	6.835	6.434	6.073	5.451	4.937	4.507	4.143	3.981	3.831	3.562	3.327	2.855	2.499	2.222	2.000
25.......	22.023	19.523	15.622	12.783	10.675	9.077	7.843	6.873	6.464	6.097	5.467	4.948	4.514	4.147	3.985	3.834	3.564	3.329	2.856	2.499	2.222	2.000
26.......	22.795	20.121	15.983	13.003	10.810	9.161	7.896	6.906	6.491	6.118	5.480	4.956	4.520	4.151	3.988	3.837	3.566	3.330	2.856	2.500	2.222	2.000
27.......	23.560	20.707	16.330	13.211	10.935	9.237	7.943	6.935	6.514	6.136	5.492	4.964	4.524	4.154	3.990	3.839	3.567	3.331	2.856	2.500	2.222	2.000
28.......	24.316	21.281	16.663	13.406	11.051	9.307	7.984	6.961	6.534	6.152	5.502	4.970	4.528	4.157	3.992	3.840	3.568	3.331	2.857	2.500	2.222	2.000
29.......	25.066	21.844	16.984	13.591	11.158	9.370	8.022	6.983	6.551	6.166	5.510	4.975	4.531	4.159	3.994	3.841	3.569	3.332	2.857	2.500	2.222	2.000
30.......	25.808	22.396	17.292	13.765	11.258	9.427	8.055	7.003	6.566	6.177	5.517	4.979	4.534	4.160	3.995	3.842	3.569	3.332	2.857	2.500	2.222	2.000
40.......	32.835	27.355	19.793	15.046	11.925	9.779	8.244	7.105	6.642	6.234	5.548	4.997	4.544	4.166	3.999	3.846	3.571	3.333	2.857	2.500	2.222	2.000
50.......	39.196	31.424	21.482	15.762	12.234	9.915	8.304	7.133	6.661	6.246	5.554	4.999	4.545	4.167	4.000	3.846	3.571	3.333	2.857	2.500	2.222	2.000

Figure 12.5 Present value of $1 received annually for N years.

The inevitable error in projections of future benefits is considered too large to justify such refinement. In any case, use of present value tables constructed on the assumption of lump returns at the end of each year of economic life is conservative; it yields smaller present values than would the use of tables based on the assumption of continuous return.

The process of converting benefits to present values is called *discounting.*

12.9.4.2. Required Rate of Return. It's apparent that a significant factor in determining present values is the rate of return expected on an investment. The rate used by an organization should be the minimum rate that it can stand. This is called the *required rate of return.*

Theoretically, the required rate of return should be set at the cost the organization experiences to raise capital. In practice, this cost is difficult to fix. Consequently most organizations settle on a rate, such as 10 percent or 20 percent, that approximates this cost of capital. In this book we use a required rate of return of 15 percent.

12.9.4.3. Net Present Value. To arrive at the net present value of an investment, you reduce the present value of the investment's benefits by the development costs of the investment. Thus the net present value of our example investment A is the present value of its benefits, $67,060, reduced by its development costs, $40,000, for a difference of $27,060. This demonstrates that investment A will return a profit. In general, any investment with a positive net present value will return a profit.

Theoretically, as long as a firm can find investments with positive net present values, it should continue to raise funds at its required rate of return and make the investment. The required rate of return is set at the minimum so that the net present value of investments exhibits this characteristic.

As illustrated in Figure 12.2, an investment may require the expenditure of development costs over a period of years. Theoretically, discounting future development costs makes as much sense as discounting benefits. Again, however, the procedure is seldom practiced; it's considered a refinement that the error inherent in projections can't justify.

12.9.4.4. Profitability Index. The net present value of an investment tells you whether the investment is profitable. This in itself is useful information, since you don't want to make unprofitable investments.

However, an organization typically has a number of profitable investment opportunities but not enough capital to take advantage of all of them. By themselves, the net present values of the investment opportunities won't allow a ranking of the opportunities by degree of profitability. For example, a $5,000 investment with a present value of $6,000 is generally considered more profitable than a $1 million investment with a present value of $1,001,000, even though each has a net present value of $1,000.

The solution to this problem is to create, for each investment, a profitability index, which is defined as the present value of the investment divided by its development costs. It should be apparent that an investment is unprofitable if its profitability index is 1 or less.

Let's see how this works out for our example investments.

Investment	A	B
Development costs	$40,000	$50,000
Annual benefit	$20,000	$30,000
Economic life	5 years	3 years

We've already determined that investment A is profitable and that its present value is $67,060. Then its profitability index is:

$$\frac{\$67{,}060}{\$40{,}000} = 1.6765$$

Using Figure 12.5, we find that the present value of investment B at a required rate of 15 percent is:

$$2.283\ (\$30{,}000) = \$68{,}490$$

Its profitability index is then

$$\frac{\$68{,}490}{\$50{,}000} = 1.3698$$

Thus A is the preferred investment.

12.9.4.5. Residual Value. Up to now we've assumed that an investment is worthless at the end of its economic life. This is often not the case. For example, the investment may be in machinery, which usually retains some value at the end of its economic life. At worst, it can be sold for scrap. The value an investment retains at the end of its economic life is called its *residual value.* When evaluating the profitability of an investment, its residual value is considered a benefit that's realized in a lump sum at the end of economic life. Thus, an investment typically has an annual benefit that it realizes over its economic life, and a one-time benefit realized at the end of economic life.

12.10. A DATA PROCESSING EXAMPLE

To determine the profitability index of a proposed data processing system, you must know the following.

1. The annual operating cost of the present data processing system. By "the present data processing system" we mean those data processing operations that will be discontinued if the proposed data processing system is installed. The annual operating cost of the present system is the cost of continuing these other data processing operations if the proposed system isn't installed.
2. The annual operating cost of the proposed data processing system.
3. The annual value of all benefits, other than savings in operating cost, to be derived by installing the proposed system.
4. The development cost of the proposed system.

The user should be able to tell you the annual operating cost of the present system. You must develop an estimate of the annual operating cost of the proposed system as part of your evaluation of alternative designs. The other benefits tend to be the so-called intangible ones, and their annual value must be developed in a way we'll describe later in this chapter. The development cost can be projected from the plans you must develop to control system development.

Here's an example determination of the profitability index of a proposed system.

The first thing you must do is determine the economic life of the proposed system. This determination is, of course, up to you. However, a not uncommon estimate is five years, since this is some approximation of the life of the average data processing system. Suppose a proposed system is to be installed at the end of next year. Then if this is year 0, the life of the proposed system might be estimated to extend from year 2 through year 6.

Suppose the present system operating cost for the next year is estimated as $100,000. Also, suppose this operating cost is estimated to increase at 10 percent a year, because of business expansion, inflation, etc. Then the projection of present system operating costs over the life of the proposed system is as follows.

Present System Operating Costs (thousands)

Year	2	3	4	5	6
	$110	121	133	146	161

Suppose that the proposed system operating cost for its first year of operation is estimated to be $75,000 and that this operating cost is also expected to increase at 10 percent a year. The projection of proposed system operating costs over its life is as follows.

Proposed System Operating Costs (thousands)

Year	2	3	4	5	6
	$75	83	91	100	110

Subtracting the proposed system operating costs from the present system operating costs gives the operating benefit of the proposed system.

Operating Benefit (thousands)

Year	2	3	4	5	6
	$110	121	133	146	161
	75	83	91	100	110
	$ 35	38	42	46	51

Suppose the value of all other benefits to be derived by installing the proposed system is estimated to be $10,000 during the first year of the proposed system's operation. Suppose further that the value of these benefits is also expected to increase at 10 percent a year. The projection of the value of all other benefits is then as follows.

All Other Benefits (thousands)

Year	2	3	4	5	6
	$10	11	12	13	15

Adding the operating benefit to the value of all other benefits gives the total value of the benefits of the proposed system.

Total Benefit Value (thousands)

Year	2	3	4	5	6
	$10	11	12	13	15
	35	38	42	46	51
	$45	49	54	59	66

The concept of residual value is applied to a data processing system as follows. At the end of its life, a proposed system will be replaced by yet another system. It may be possible to retain some part of the proposed system and incorporate it into the replacement system. Such retention of part of the proposed system reduces the development cost of the replacement system. The amount of this reduction is the residual value of the proposed system.

The residual value of the proposed system is generally estimated as some percentage of its development cost. Suppose the percentage is estimated at 20 percent and the development cost as $50,000. The estimate of the residual value of the proposed system is then $10,000.

Here's a summary of the future benefits of the proposed system, which is the total benefits plus the residual value.

Future Benefits (thousands)

Year	2	3	4	5	6	Residual
	$45	49	54	59	66	10

To convert these future benefits into a present value we use the table in Figure 12.4. Assuming a required rate of return of 15 percent, the calculation of the present value of the future benefits received from the installation of the proposed system is shown below.

Year		
2	0.756($45,000) =	$ 34,000
3	0.658($49,000) =	32,000
4	0.572($54,000) =	31,000
5	0.497($59,000) =	29,000
6	0.432($66,000) =	29,000
Residual	0.432($10,000) =	4,000
		$159,000

Thus the present value of the proposed system is $159,000.

Dividing the present value of the proposed system, $159,000, by the system development cost of $50,000 gives a profitability index of 3.18. Since this profitability index is greater than 1, the proposed system is economically feasible.

12.11. QUANTIFYING INTANGIBLES

In doing a cost benefit analysis, it is particularly difficult to estimate the value of the benefits of the proposed system other than the operating benefit. Frequently, these other benefits are referred to as intangible, which implies that they can't be quantified. If a benefit truly can't be quantified, then it can't be taken into consideration in a cost benefit analysis.

However, in many cases benefits that are initially classified as intangible are subject to a considerable amount of quantification. For example, suppose the proposed system is a personnel system and the benefit is reduced turnover. Isn't this an intangible benefit?

To demonstrate that in fact reduced turnover isn't an intangible, we must show that it is possible to do two things.

1. Estimate the extent of the reduction
2. Estimate the savings that result from not having a person leave the company

Given these two estimates we can quantify the benefit.

Making both these estimates requires the cooperation of the user. The contribution the data processing division analyst can make to the estimation process is to help the user in his quantification efforts.

With respect to estimating the extent of turnover reduction, the basic question is, By what percentage will turnover be reduced? The user may be unable or unwilling to propose a specific percentage. However, he may be willing to say that there's a 20 percent chance that turnover will be reduced by 5 percent, a 50 percent chance that it will be reduced by 4 percent, and a 30 percent chance that it will be reduced by 3 percent. And this is all that's really needed, because as the following calculation shows, what the user is really saying is that his best guess is that turnover will be reduced by 3.9%.

$$\begin{aligned} 0.2(0.05) &= 0.01 \\ 0.5(0.04) &= 0.02 \\ 0.3(0.03) &= \underline{0.009} \\ & 0.039 \end{aligned}$$

This percentage can be converted to an absolute figure as follows. If the company's turnover rate is presently 100 people a year, then we can expect the proposed system to reduce this figure by 3.9 people a year.

Now, how much will this turnover reduction save? To answer this question, the user and the analyst must isolate the costs associated with one person leaving the company. The results of this investigation might be as follows.

1. Termination pay	$ 600
2. Recruitment costs to hire a replacement	1,200
3. Replacement's salary and overhead while being trained	7,200
4. Training costs	1,200
	$10,200

Thus the annual value of the benefit of reduced turnover is

$$3.9(\$10,200) = \$39,780.$$

12.12. SIGNIFICANCE OF CASH FLOW

Determining that the profitability index of a proposed data processing system is greater than 1 is analogous to determining that it's desirable to be on the other side of the river. However, if you drown crossing the river, the desirability of being on the other side is academic. A cash flow analysis is analogous to determining the depth of the river at all points in the crossing. This point is most graphically demonstrated by the chart in Figure 12.1, which could be considered a cross section of a river bed. If you can't afford to go into debt to the extent indicated by the cash flow analysis, then the ultimate benefits of the undertaking are evanescent. For example, in the situation described in Figure 12.2, if you can't afford a cash drain of $136,000, then any question of the ultimate benefit of the system is academic. Consequently, a complete cost benefit analysis consists of a profitability index and a cash flow analysis.

If the form in Figure 12.2 were applied to a data processing system, receipts would be the sum of:

1. The operating costs of the present system
2. Other benefits derived by installing the proposed system

EXERCISES

1. List the steps taken to develop a profitability index for a proposed data processing system.
2. What is the present data processing system?
3. What is the operating cost of the present data processing system?
4. What is the residual value of a data processing system?
5. What is the present value of a future benefit?
6. What is the required rate of return?
7. How would you go about quantifying the benefit "reduction in accident rate"?
8. Given the following statistics, develop a profitability index.
   a. The proposed data processing system is to be installed three years from now. Its economic life is four years.
   b. The operating cost of the present data processing system is estimated to be $500,000 for the next year.
   c. The operating cost of the proposed system during its first year of operation is estimated to be $400,000.
   d. The value of all benefits, other than operating cost benefits, to be realized by installing the proposed system, is estimated to be $200,000 per year.
   e. The operating cost of the present system, the operating cost of the proposed system, and all benefits other than operating cost benefits are expected to increase at a rate of 10 percent per year.
   f. It's estimated that 10 percent of the proposed system will be incorporated into its replacement.
   g. The required rate of return is 15 percent.
   h. The cost to develop the system is estimated to be $500,000.
9. Is the system described in Exercise 8 economically feasible? Why?
10. Do a cash flow analysis by year of the proposed system in Exercise 8. Assume that $100,000 of the development costs will be spent in year 1, and $200,000 in each of years 2 and 3.

Answers

Chapter 2

1. The faster you read, the greater your comprehension. The explanation for this relationship is that the faster you read, the less chance your mind has to wander; therefore your concentration on what you're reading, and consequently, your comprehension of the material, is increased.

2. Regression is returning to material already read for a second look.

3. Fixation is focusing your eyes on a particular point of the printed page. Fixation is necessary for reading, since your eyes can't focus adequately for reading while they're moving. However, fixation can contribute to decreased reading speed, because we have a tendency to stare at words longer than necessary to read them.

4. Vocalization is the pronunciation of words as we read them.

5. Vocalization decreases reading speed because our eyes can read words faster than our lips can pronounce them.

6. The vocalization habit can be broken by preventing vocalization from occurring. The simplest way to do this is to chew gum while you read.

7. Pinpoint perception is seeing only what your eyes are pointed at.

8. Pinpoint perception decreases reading speed by increasing the number of fixations required to read a given amount of printed matter. It reduces comprehension because your mind grasps information in phrases, and pinpoint perception transmits information to your mind one word at a time.

9. Pinpoint perception is a psychological limitation, because you actually have a span of perception—you can see a considerable amount around the point at which your eyes are focused.

10. The alternative to pinpoint perception is to use your span of perception to read a phrase with each fixation.

11. Given an article you're going to read, the first thing you should do is orient yourself to the article—determine in your mind what questions you expect the article to answer. If you're familiar with the subject addressed by the article, you should be able to formulate these questions out of your own knowledge. In many cases, you'll already have the questions formulated. They'll be ones to which you don't have answers, and you'll be reading the article to see whether it answers these questions. If you're not familiar with the article's subject, skim the article before reading it; look at the abstract, section titles, topic sentences, and summary. Then formulate your questions on the basis of your skim.

12. Scanning is the skipping of words, phrases, sentences, paragraphs, sections, and even whole chapters of the material being read. Scanning is justified when you're oriented to the reading matter—when you know what questions you want it to answer. If scanning gives you the answers you want or demonstrates that the material doesn't contain these answers, you've lost nothing by not reading those parts of the material that you've skipped.

Chapter 3

1. You want to listen to someone when you suspect he has information bearing on your ability to attain your goals.

2. You develop the proper listening attitude by:

a. Remaining amenable to the idea that you may have to change your ideas of how things are

b. Treating the other person as your equal

c. Resisting the tendency to let your emotional reaction to words inhibit your reception of information

3. You get the other person to tell you what you need to know by:

a. Giving him priority

b. Setting him at ease

c. Removing distractions

d. Using nondirective conversation

e. Letting the other person know that you're listening

f. Withholding judgment until he's said everything he wants to say

g. Helping him out

h. Taking action

i. Taking notes

4. You can actively contribute to a conversation while you're listening by looking, in the other person's words, for:

a. The principles he endorses

b. The facts he marshals to support these principles

5. You can determine these principles and facts by:

a. Being prepared on the subject to be discussed

b. Anticipating the other person
c. Evaluating what you hear
d. Summarizing

6. When listening, you should use your spare thinking time to:
a. Anticipate what the other person is going to say
b. Review what he's said
c. Analyze the conversation

Chapter 4

1. Interviewing is collecting information from another person through conversation.

2. The first problem the interviewer faces is getting the respondent to participate willingly in the interview. The interviewer solves this problem by orienting the respondent to the interview—that is, the interviewer tells the respondent:
a. What procedures the interviewer is working to improve
b. How the respondent can help by supplying information
c. What information the interviewer needs
d. How the respondent is to supply this information

3. The interviewer must convince the respondent that:
a. The two of them have an overlap of knowledge in the area under discussion.
b. The interviewer is capable of constructively using the information supplied.
c. The interviewer respects the respondent's ability to provide the information needed.
d. The respondent is free to express himself without fear of being judged by the interviewer.

4. The interviewer starts the questioning process with the most general question in the subject area to give the respondent maximum opportunity to communicate the needed information.

5. The best way for the interviewer to prepare for an interview is to develop an interview guide.

6. If the job is to interview a programmer about the procedure she follows in doing her job, the interview guide might look as follows.
a. What are the steps you take in the preparation of a program?
b. How do you schedule your work?
c. What do you do to keep track of whether you're staying on schedule?
d. What precautions do you take to see that the program you prepare does what the system designer expects it to do?
e. How are specifications for a program prepared? If the programmer prepares them, when is this done?
f. What kind of a logical analysis is done?
g. When is the logical analysis done?
h. At what level is the logical analysis done?
i. When is the modular organization of the program done?
j. When are notes incorporated into the code?

k. Are meaningful data and paragraph names used?
l. What kind of a test plan is developed?
m. How is test data obtained?
n. Are test data results predetermined?

7. Two ways in which the interviewer can get the respondent to expand on his remarks are:
a. To summarize
b. To use a nondirective probe

8. Three types of nondirective probes are:
a. A brief assertion of interest
b. A pause
c. A neutral phrase

9. Three scales for rating the effectiveness of a probe are:
a. Acceptance—does the probe contribute to the interviewer-respondent relationship?
b. Validity—does it leave the respondent free to expand on his comments as he sees fit?
c. Relevance—does it lead the respondent toward the information for which the interviewer is looking?

10. a. I: I can certainly understand why you'd *want* sales to increase next year. Now, how about your *expectations*? Regardless of what you *want* to happen, what do you really think is *going* to happen?

Here the respondent simply failed to answer the question. Instead of saying what he thought was going to happen, he described what he hoped would happen. The interviewer politely clarifies the distinction for the respondent and repeats the question.

b. I: I see. In what ways is it OK?

Here the respondent didn't supply enough detail in response to the interviewer's question. The interviewer gives the respondent an opportunity to expand on his answer. The respondent's "I guess" may indicate that some aspects of the new system aren't satisfactory to him, and if he doesn't voice any complaints in his subsequent comments, the interviewer should ask him whether there was any significance to the qualifier.

c. I: And these people whom we do employ: Do you think each one of them will turn out the same amount of work next year as he does now, or will he turn out more each passing year?

Here the interviewer rewords the question so as to avoid the irrelevancy introduced by the respondent's first answer.

d. I: To what would your people object?

Here the respondent is confused as to what his position is. The interviewer invites the respondent to clarify his thinking.

11. Threatening questions should be the last ones in an interview because:
a. If the respondent reacts negatively, the interviewer will at least have gotten the respondent's cooperation during the rest of the interview.
b. The interviewer has had the maximum opportunity to gain the respondent's confidence, after which the question may no longer appear threatening to the respondent.

12. When terminating an interview, the interviewer must:

a. Give the respondent a last opportunity to mention anything he thinks is significant.

b. Get the respondent's permission to return with subsequent questions.

c. Get the respondent's commitment to review the interview report.

Chapter 5

1. The purposes of business communications are to convey information and to arrive at decisions. Reports are used to convey information, meetings to arrive at decisions.

2. If you have some information to convey, you should write a report. At best, people retain only part of what they hear at a meeting.

3. If you want Charlie to make a decision, you should meet with him. Decisions require an interchange of ideas and attitudes.

4. If a decision must be made and preparation for making the decision requires absorbing much information, then you should write a report to convey the information and follow up with a meeting in which the decision is to be made.

5. A meeting begins with a problem statement, which is made in terms of what exists versus what the meeting members want to exist. The next step is to establish the area of common agreement between you and the meeting members. The third step is a presentation that shows how your proposed solution to the problem flows logically from the area of common agreement. The fourth step is to get the meeting members to endorse your proposed solution. The last step is to outline the steps to be taken as a result of the agreement.

6. If you don't know the answer to a question, admit it, and tactfully ask the audience whether they know the answer.

7. Every good presentation ends with a proposal of what you want the audience to do.

8. The three essential parts of a good presentation are:

a. Problem statement

b. Development

c. Problem solution

9. The essence of a problem statement is what exists versus what the audience wants to exist.

10. You should encourage your audience to make observations and ask questions, because when they do, it proves that they're paying attention.

11. Every good visual aid should have a heading.

12. A visual aid should be big enough to be seen.

13. The best way to make a point is by example.

14. You avoid jargon by telling your audience only the technical information they must know and by expressing this information in terms they understand, either by analogy or by definition.

15. The development of every good presentation must begin with the attitudes and beliefs of your audience.

16. To make a long presentation more effective, organize a team to make the presentation.

Chapter 6

1. A set of functional specifications is the product of the functional specification activity.

2. The three parts of the functional specifications are:
   a. System description
   b. Acceptance criteria
   c. Measuring benefits

3. The five parts of the system description are:
   a. Output
   b. Records
   c. Input
   d. Processing
   e. Control

4. The output section describes:
   a. Content
   b. Presentation method
   c. Estimated volume
   d. Time schedule

5. The input section describes:
   a. Content
   b. Entering method
   c. Estimated volume
   d. Time restraints

6. The description in the processing section is carried out to the extent necessary to define the origin of each output and record field.

7. The processing section doesn't need to be written from scratch because most of the required information can be specified by reference to other documents.

8. A comprehensive set of test data examples makes the processing section most meaningful to the user.

9. The acceptance criteria section specifies "who" and "how" for the preparation of inputs for and review of the outputs of the acceptance demonstration.

10. The measuring benefits section specifies how the benefits of the system are to be measured to determine whether they measure up to the original claims.

11. a. This statement doesn't belong in the functional specifications. It tells *how* the information is to be obtained.

b. Neither does this statement belong in the functional specifications. Once again, it tells *how* information is to be collected.

c. This statement is part of the functional specifications. It tells *what* the contents of the records are.

d. Not part of the functional specifications. The statement says *how* something is going to be done.

e. Part of the functional specifications. This is a definition of the overtime pay rate. This is an interesting example, because it could be maintained that this statement tells how to calculate the overtime pay rate. However, this statement defines what the overtime pay rate is and represents something to which the user and the data processing division must agree before agreeing on the functional specifications. As to how overtime pay is actually computed, when it comes time to make this calculation, the programmer may decide to increase overtime hours by one half and then multiply by the regular pay rate, or the programmer may decide on some other technique. So you can see that this statement specifies *what* is to be done, not *how* it's to be done.

f. No. This statement refers to *how* records are to be stored.

g. No. This statement refers to hardware requirements.

h. Yes. This states *what* must be done—specifically, under what time restraints the output must be delivered.

i. Yes. This is the same as the previous statement, except that this statement refers to a realtime output, while the previous one refers to a batch output.

j. No. This statement refers to *how* records are to be organized.

Chapter 7

1. The best place for instructions on how to fill out a form is on the form itself.

2. You can tell whether you've allowed enough space on a form for the entry of the requested information by trying out the form in a variety of circumstances in which it will be used.

3. Three common failures in the specification of form content are:
   a. Not all required information is included
   b. Superfluous information is included
   c. An insufficient number of copies is specified

4. The requested layouts are as follows.
   a. Applicant's name and address (Figure A7.4a)
   b. Applicant's sex (Figure A7.4b)
   c. The date on which an order is made (Figure A7.4c)
   d. The amount of a check (Figure A7.4d)

5. Two methods of avoiding the need to supply the same information on forms more than once are:
   a. Combine related forms
   b. Preprint information

APPLICANT

NAME (FIRST, MIDDLE INITIAL, LAST)		
STREET (NUMBER, NAME)		
CITY	STATE	ZIP CODE

Figure A7.4a

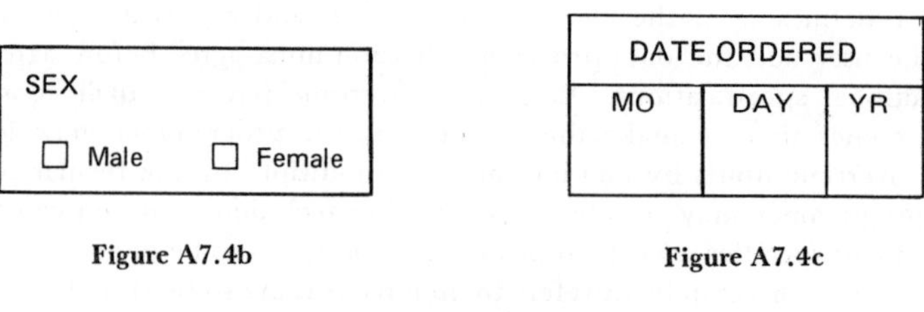

Figure A7.4b

Figure A7.4c

CHECK AMOUNT	
DOLLARS	CENTS

Figure A7.4d

6. One method of facilitating the routing of copies of a multipart form is to make each copy a different color.

7. The preferred approach to selective reproduction of information on a multipart form is die-cut carbons.

8. If a form is laid out on both sides of a sheet of paper, the front of the form must contain a prominently placed notice that both sides of the form are to be completed.

9. The three principal factors to be considered when determining form layout are:

a. The ease with which the form is completed
b. The ease with which it's used after it's completed
c. Printing requirements

10. The layout considerations imposed by the requirement to complete a form on a typewriter are as follows.

a. Horizontal spacing must conform to the rachet spacing of the typewriter. Spacing should be either all single-spaced or all double-spaced.
b. Standardize tab stops.
c. Avoid the necessity for the typist to back up the platen.
d. Minimize carriage returns and skips.
e. Make it easy for the typist to align the form in the typewriter.

11. The user is responsible for form content.

12. The layout considerations imposed by the requirement to bind a form in a binder are as follows.

a. Leave sufficient margin for binding so the form can be read after it's in the binder.
b. Put the identifying information on the form where it can be found by just flipping the forms in the binder.

13. You can be sure you've laid out a form so that the blank form can be reproduced by checking with the person who's going to do the printing for you.

14. Before incorporating a suggested change into a form currently being used, you should check to see that the change is agreeable with all the form users.

15. Two reasons why all forms should be controlled from a central point are:

a. To avoid duplication of forms

b. To enforce conformity with form design standards

16. If the same information is to appear on more than one form, you'd interlock the forms by laying out the common information the same way on all the forms.

17. The purpose of an input form is to introduce *new* data into the system.

18. The primary consideration in input form layout is ease of completion.

19. The best form size is 8-1/2 by 11, because it's the most standard size and is therefore most easily bound and filed.

20. If instructions for filling out a form are physically separated from the form itself, you relate the instructions to the form by numbering the entries on the form and keying the instructions to the numbers.

21. The form designer is responsible for form layout.

22. Figure A7.22 shows a possible waitress check layout. The checks are supplied in the form of a pad. There are many alternative layouts that will suit the purpose as well. However, these layouts should all take into consideration the following.

a. Since the restaurant's menu is highly structured, it would be a mistake to use a relatively free-form waitress check. Instead, the check should be structured to help the waitress in filling it out.

b. The waitress check is to be filled out by hand. Therefore the space for entering information shouldn't be cramped. Also, the check should either be strong enough itself to make writing easy, or the checks should be supplied in the form of a pad that can provide this strength.

c. The waitress will have to carry a number of checks around with her, and she needs her hands free for other purposes. Therefore, the checks must be small enough to fit into a pocket. Having to remove the checks from and insert the checks into her pocket also argues for supplying them in the form of a pad.

d. Since the waitresses are collecting the money, this collection must be controlled. This is done by controlling the checks themselves by prenumbering them. While the exercise doesn't specify this, your clue is that the customer pays the waitress.

23. A possible form layout is shown in Figure A7.23. A procedure for completing and handling the form follows.

Procedure for Using the Request for Change Form

The *Request for Change Form* is initiated by the user. It is then submitted to the project leader, who acknowledges the receipt of and files for future action all postponable changes. For changes requiring immediate attention, the project leader spells out the impact of the change on the system and the project and then estimates the change in schedule and budget required to institute the change in specifications. This information is transmitted to the user. When the user approves the changes in schedule and budget, the specifications change is implemented.

The *Request for Change Form,* a four-part form, is completed in the following way.

1. The requesting department is identified.

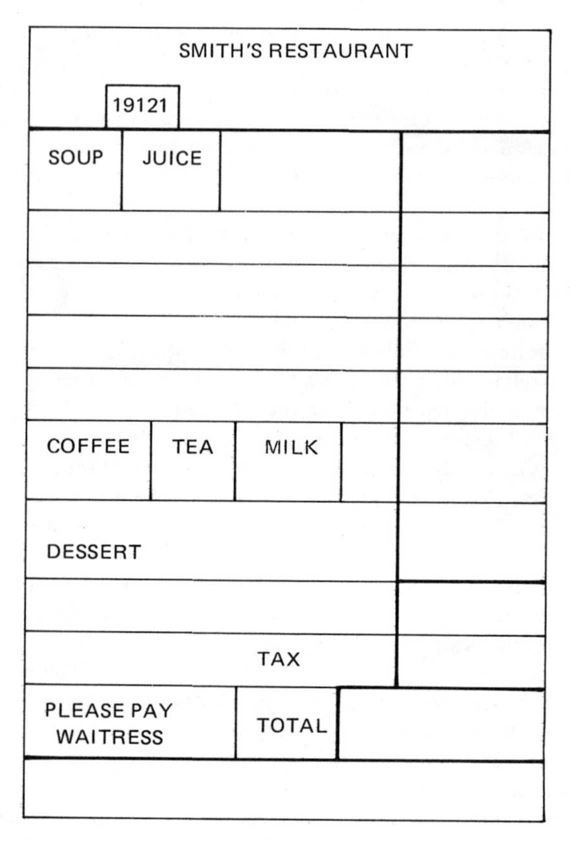
SMITH'S RESTAURANT

19121				
SOUP	JUICE			
COFFEE	TEA	MILK		
DESSERT				
		TAX		
PLEASE PAY WAITRESS	TOTAL			

Figure A7.22

2. The project is identified by project number.
3. The requested change is described.
4. The benefits to be derived from the change are listed.
5. The change is classified as:
   a. Immediate—a change that must be immediately incorporated into the design
   b. Postponable—a change that is desirable but can wait for implementation until the initial system is accepted
6. The form is approved by the requesting department head.
7. The original and two copies are submitted to the systems and programming group.
8. If the change is classified as postponable, the project leader acknowledges the request by signing the form and returns the copies to the user.
9. If the change is classified as immediate:
   a. The project leader spells out the impact of the change.
   b. The estimated change in schedule is entered.
   c. The estimated change in budget is entered.

REQUEST FOR CHANGE FORM

<table>
<tr><td colspan="4">REQUESTING DEPT
Payroll</td><td>PROJECT NO.
12345</td></tr>
<tr><td colspan="5">DESCRIPTION OF CHANGE
Modify withholding to continue the surtax deduction.</td></tr>
<tr><td colspan="5">BENEFITS
Conform to governmental regulations.</td></tr>
<tr><td>X</td><td>IMMEDIATE</td><td></td><td colspan="2">POSTPONABLE</td></tr>
<tr><td colspan="4">DEPARTMENT HEAD</td><td>DATE
9/3/69</td></tr>
<tr><td colspan="5">IMPACT
Addition of a routine to the Detail Calculate Program.</td></tr>
<tr><td colspan="2">DELAY IN SCHEDULE (IN WEEKS)
one week</td><td colspan="3">BUDGET INCREASE (IN DOLLARS)
$450</td></tr>
<tr><td colspan="4">PROJECT LEADER</td><td>DATE
9/16/69</td></tr>
<tr><td colspan="4">DEPARTMENT HEAD</td><td>DATE
9/25/69</td></tr>
</table>

(USE MULTIPLE FORMS AS REQUIRED)

Figure A7.23

d. The project leader signs the form.
e. The original and one copy are returned to the user.
f. If the user accepts the schedule and budget changes, the form is approved by the requesting department head.
g. The original is submitted to the systems and programming group.

The form should be multipart. The parts could be different colors to identify routing, in which case this should be spelled out in the procedure. The procedure could be printed on the back of the form, in which case a notice to this effect should appear on the front of the form.

Two points are of importance.

a. This form could be filled out either by hand or with a typewriter. Therefore it must be laid out for use in either way.
b. The procedure is an integral part of the form. A form of this type without an accompanying procedure is incompletely designed.

Other common mistakes in laying out this form are as follows.

a. Not leaving a margin on the left for binding purposes.
b. Not making the form 8-1/2 by 11 for ease of storage.
c. Not leaving a margin at the bottom of the form. (If the form requires information to be entered at the bottom and if the form is completed on a typewriter, there won't be sufficient paper left for the typewriter to grasp while the bottom of the form is being completed.)
d. Requiring too many tab stops on the typewriter.

It could be argued that the school solution is faulty in that the form and the instructions for completing it are physically separate, yet the form entries aren't numbered and the instructions keyed to these numbers. The rationale for the school solution is that:

1. There are a minimum number of entries on the form, and the form is always completed from top to bottom with no skipping of entries.
2. Each entry is identified by a caption.
3. The instructions are basically keyed by caption.

24. A possible check layout is shown in Figure A7.24. Principles incorporated into this layout are as follows.

a. The check fits into a standard business envelope. (Many people like to bank by mail.)

b. The stub is the same size as the check. Consequently the stub can be folded on the perforation over onto the check, and the result is an easily handled package.

c. The total form has the same width as standard high-speed printer pap. Therefore no tractor adjustment is normally necessary to insert the form in the printer.

d. The normal check requires only two lines of print.

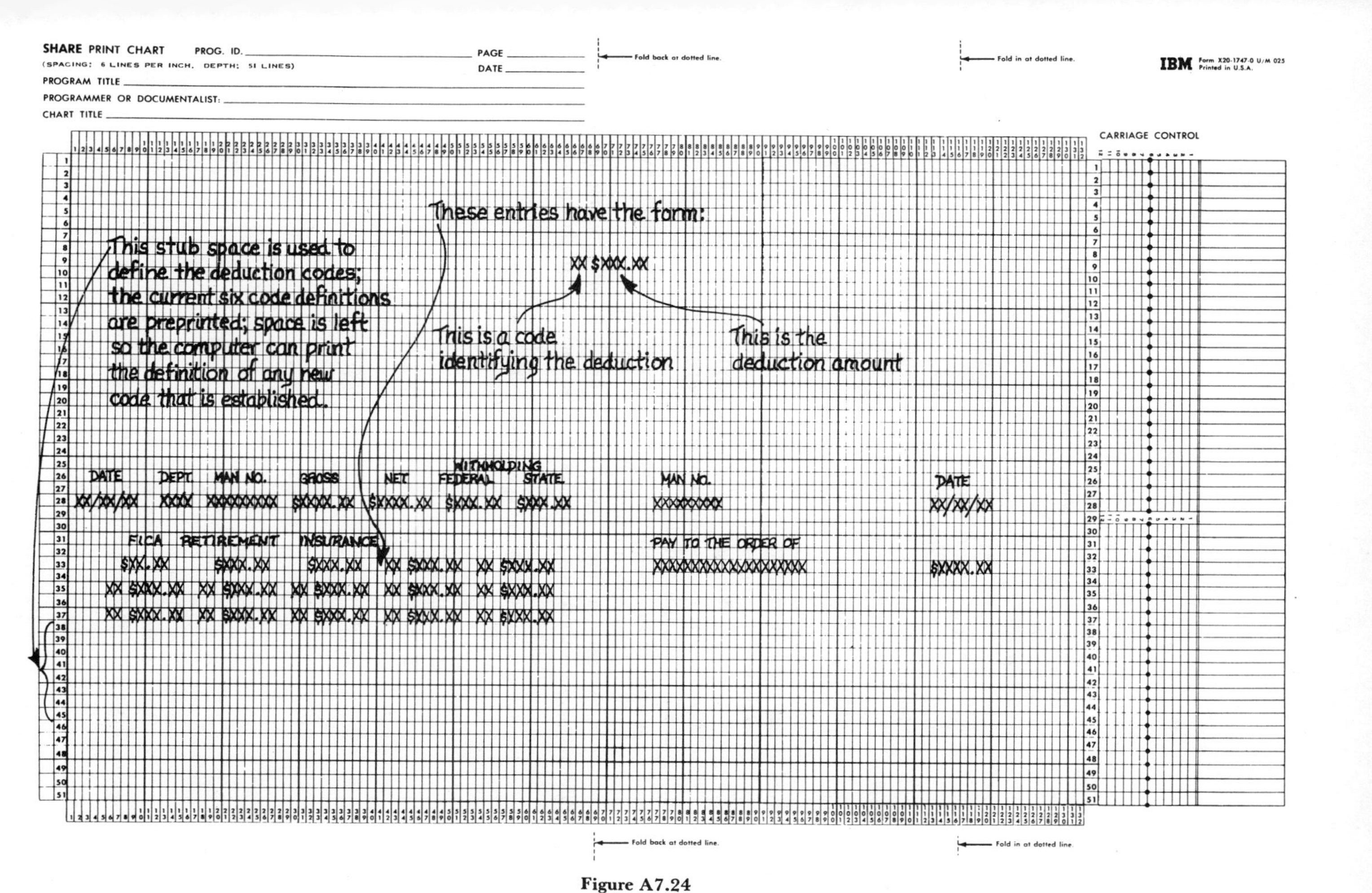

SHARE PRINT CHART
PROG. ID.
PAGE
DATE
(SPACING: 6 LINES PER INCH, DEPTH: 51 LINES)
PROGRAM TITLE
PROGRAMMER OR DOCUMENTALIST:
CHART TITLE
Fold back at dotted line.
Fold in at dotted line.
IBM
Form X20-1747-0 U/M 025
Printed in U.S.A.
CARRIAGE CONTROL
This stub space is used to define the deduction codes; the current six code definitions are preprinted; space is left so the computer can print the definition of any new code that is established.
These entries have the form:
XX $XXX.XX
This is a code identifying the deduction
This is the deduction amount
WITHHOLDING
DATE DEPT. MAN NO. GROSS NET FEDERAL STATE MAN NO. DATE
XX/XX/XX XXXX XXXXXXXX $XXXX.XX $XXXX.XX $XXX.XX $XXX.XX XXXXXXXX XX/XX/XX
FICA RETIREMENT INSURANCE PAY TO THE ORDER OF
$XX.XX $XXX.XX $XXX.XX XX $XXX.XX XX $XXX.XX XXXXXXXXXXXXXXXXXX $XXXX.XX
XX $XXX.XX XX $XXX.XX XX $XXX.XX XX $XXX.XX XX $XXX.XX
XX $XXX.XX XX $XXX.XX XX $XXX.XX XX $XXX.XX XX $XXX.XX

Figure A7.24

Chapter 8

1. The object of terminal dialogue design is to avoid frustrating the operator.

2. Seven goals of good dialogue design are:
   a. Communication
   b. Minimum operation action
   c. Low requirement of special skills
   d. Standardized operation
   e. Stability
   f. Satisfactory response time
   g. One-hand operation

3. Standard terminals are better because customizing terminals for the first design of the initial application generally reduces the effectiveness of:
   a. Redesigns
   b. Subsequent applications

4. Two things that can be expected of a dedicated operator, as distinguished from a casual operator, are:
   a. Skill in terminal use
   b. Greater willingness to undergo a lengthy training program

5. System-structured dialogues are system initiated; operator-structured dialogues are operator initiated.
   a. System-structured dialogues handle operator decisions by means of menu selection. Operator-structured dialogues use action codes.
   b. System-structured dialogues handle data entry by means of form filling. Operator-structured dialogues precede each unit of data with an action code.

6. Thirteen operations characteristic of dialogues are:
   a. START
   b. HELP
   c. LOCATION RETURN
   d. Field formating
   e. RELEASE
   f. Cursor movement
   g. CANCEL
   h. Frame movement
   i. DELETE
   j. INSERT
   k. INTERRUPT
   l. RESUME
   m. TERMINATE

7. The operator should be notified of an error by an audible signal so he doesn't have to be on the lookout for error warnings on the display.

8. The maximum acceptable variation in system response time is a standard deviation less than or equal to half of average response time.

9. The maximum average response time for:
   a. A START operation is one second
   b. An inquiry by an operator talking to a customer is two seconds
   c. A system response in a conversation is two seconds

10. The general approach to maintaining quality dialogue without inordinately increasing network cost is distribution of the intelligence throughout the network.

11. A dialogue can be considered specified when it has passed all its simulation requirements, both manual and automated.

12. The problem posed is that the operator is casual, but the nature of the dialogue requires program-like expressions. The solution is the information center.

13. a. The frames are shown in Figures A8.13a, A8.13b and A8.13c.

b. Frame 1 follows the START operation. Frame 2 follows frame 1. Frame 2 follows itself. To get out of this loop, the operator must take positive action, such as depressing a key, which causes frame 3 to be displayed. Frame 1 follows frame 3.

c. Validity checks are as follows.
Invoice number—mandatory field
Month—numeric field, limit, mandatory field
Day—numeric field, limit, mandatory field
Year—numeric field, mandatory field
Sold to Name—alphabetic field, mandatory field
Sold to Street—none
Sold to City—alphabetic field, mandatory field
Sold to State—alphabetic field, mandatory field
Sold to Zip Code—numeric field
Ship to Name—alphabetic field
Ship to Street—none
Ship to City—alphabetic field
Ship to State—alphabetic field
Ship to Zip Code—numeric field
Stock Number—valid value, mandatory field
Description—none
Quan—numeric field, mandatory field
Price—numeric field, mandatory field
Amount—numeric field
Total—numeric field

d. General operations of convenience to the operator in this application would be:

1. Left-justify and zero fill for numeric fields
2. Right-justify and space fill for alphabetic fields
3. Duplicate for when Ship To is the same as Sold To
4. Skip and space fill for when a nonmandatory field is absent

```
(INVOICE NUMBER)           _____
(DATE)                     __/__/__
SOLD TO:
(NAME)      ____________________________
(STREET)    ____________________________
(CITY)      _______________     (STATE)__
(ZIP CODE)_____
SHIP TO:
(NAME)      ____________________________
(STREET)    ____________________________
(CITY)      _______________     (STATE)__
(ZIP CODE)_____
```

Figure A8.13a Frame 1.

```
STOCK
NUMBER                        DESCRIPTION

_____                         ____________

QUAN          PRICE           AMOUNT

___           _____.__        _________.__
```

Figure A8.13b Frame 2.

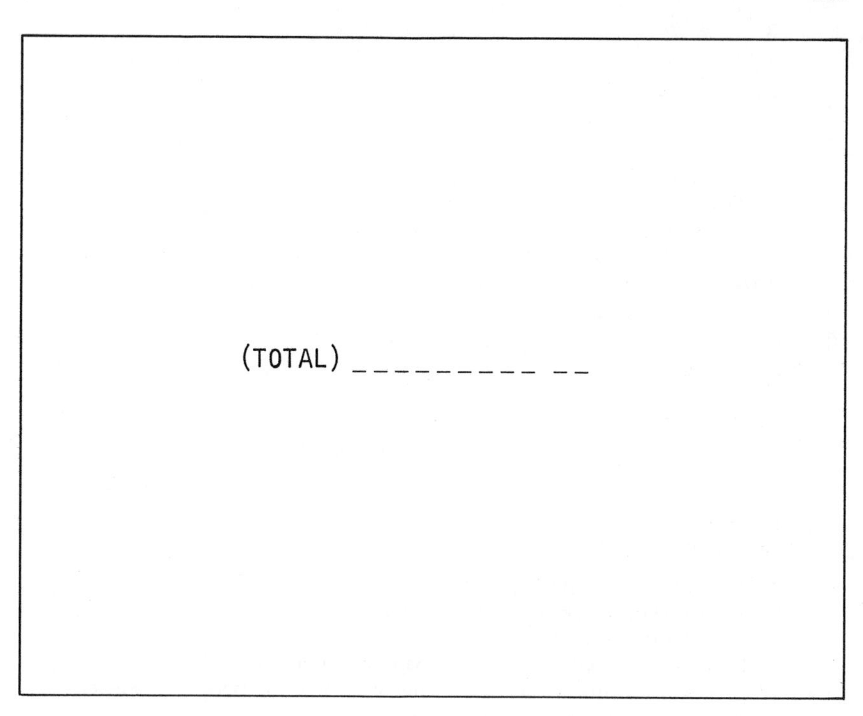

Figure A8.13c Frame 3.

Chapter 9

1. Here's how we arrived at our solution, which is shown in Figure A9.1a. The list of conditions and actions we isolated from the narrative description are shown in Figure A9.1b. These conditions and actions, stated in standardized language and with duplications removed, are shown in Figure A9.1c. Our first cut at a decision table is shown in Figure A9.1d. The excerpts from the narrative description that correspond to the rules in the table in Figure A9.1d are as follows.

1. When the market in general is going up, you should buy the stock if it is appreciating faster than the market as a whole.

2. Otherwise, you should look for another stock.

3. If you have knowledge of a special situation, check the stock's recent price and volume action. If price has been increasing on expanding volume and decreasing on contracting volume, buy the stock.

4, 5. Otherwise, your information is probably faulty, and you should hold back.

6. If the market is going up as a whole, you have knowledge of a special situation, and the stock is behaving in such a fashion as to indicate that it should be bought, buy the stock on margin.

	1	2	3	4	5	6	ELSE
MARKET APPRECIATING	Y	Y	Y	Y	Y	N	
STOCK APPRECIATING FASTER THAN MARKET	Y	Y	Y	Y	N	–	
SPECIAL SITUATION	Y	Y	Y	N	–	Y	
PRICE INCREASING ON EXPANDING VOLUME	Y	Y	N	–	–	Y	
PRICE DECREASING ON CONTRACTING VOLUME	Y	N	–	–	–	Y	
BUY THE STOCK ON A REGULAR BASIS	–	X	X	X	–	X	–
BUY THE STOCK ON MARGIN	X	–	–	–	–	–	–
LOOK FOR ANOTHER STOCK	–	–	–	–	X	–	–
HOLD BACK	–	–	–	–	–	–	X
EXIT	X	X	X	X	X	X	X

Figure A9.1a Decision table.

Conditions

1. The market as a whole is appreciating.
2. You have knowledge of a special situation.
3. The market in general is going up.
4. It (the stock) is appreciating faster than the market as a whole.
5. Otherwise (the stock is not appreciating faster than the market as a whole).
6. You have knowledge of a special situation.
7. (The stock) price has been increasing on expanding volume.
8. (The stock price has been) decreasing on contracting volume.
9. Otherwise
   (a. The stock price has been decreasing on expanding volume.
   b. The stock price has been increasing on contracting volume.)
10. The market is going up as a whole.
11. You have knowledge of a special situation.
12. The stock is behaving in such a fashion as to indicate it should be bought
    (a. The stock is appreciating faster than the market as a whole.
    b. The stock price has been increasing on expanding volume.
    c. The stock price has been decreasing on contracting volume.)
13. Otherwise
    (a. You do not have knowledge of a special situation.
    b. The stock price has been increasing on expanding volume.
    c. The stock price has been decreasing on contracting volume.)
14. The market is going up.

Actions

1. Buy a particular stock.
2. Buy the stock.
3. Look for another stock.
4. Buy the stock.
5. Hold back.
6. Buy the stock on margin.

Figure A9.1b Conditions and actions.

Conditions

1. Market appreciating
2. Special situation
4. Stock appreciating faster than market
7. Price increasing on expanding volume
8. Price decreasing on contracting volume

Actions

1. Buy the stock on a regular basis
3. Look for another stock
5. Hold back
6. Buy the stock on margin

Figure A9.1c Duplication removed.

	1	2	3	4	5	6	ELSE
MARKET APPRECIATING	Y	Y	–	–	–	Y	
SPECIAL SITUATION	–	–	Y	Y	Y	Y	
STOCK APPRECIATING FASTER THAN MARKET	Y	N	–	–	–	Y	
PRICE INCREASING ON EXPANDING VOLUME	–	–	Y	Y	N	Y	
PRICE DECREASING ON CONTRACTING VOLUME	–	–	Y	N	–	Y	
BUY THE STOCK ON A REGULAR BASIS	X	–	X	–	–	–	–
LOOK FOR ANOTHER STOCK	–	X	–	–	–	–	–
HOLD BACK	–	–	–	X	X	–	X
BUY THE STOCK ON MARGIN	–	–	–	–	–	X	–
EXIT	X	X	X	X	X	X	X

Figure A9.1d First cut.

	1	2	3	6	ELSE
MARKET APPRECIATING	Y	Y	–	Y	
SPECIAL SITUATION	–	–	Y	Y	
STOCK APPRECIATING FASTER THAN MARKET	Y	N	–	Y	
PRICE INCREASING ON EXPANDING VOLUME	–	–	Y	Y	
PRICE DECREASING ON CONTRACTING VOLUME	–	–	Y	Y	
BUY THE STOCK ON A REGULAR BASIS	X	–	X	–	–
LOOK FOR ANOTHER STOCK	–	X	–	–	–
HOLD BACK	–	–	–	–	X
BUY THE STOCK ON MARGIN	–	–	–	X	–
EXIT	X	X	X	X	X

Figure A9.1e Removing rules 4 and 5.

The narrative description implies that it has stated all the rules under which stock should be bought. Therefore under any other conditions we should hold back. That gives us our else rule.

Since rules 4 and 5 have the same actions as the else rule, they don't have to appear in our decision table. Removing them produces the decision table shown in Figure A9.1e.

In the table in Figure A9.1e, rules 1 and 3 aren't independent of rule 6. Therefore these rules are contradictory. We have to clarify our thinking.

Look at the last sentence in the narrative description. It says that unless we can answer yes to all five conditions (this is what "otherwise" must mean), then when the market is appreciating, we shouldn't worry about special situations. This means that except for the situation called for by rule 6, we should be concerned about special situations only when the market isn't appreciating. Therefore we add this crucial no entry to the first condition for rule 3, as shown in Figure A9.1f.

	1	2	3	6	ELSE
MARKET APPRECIATING	Y	Y	N	Y	
SPECIAL SITUATION	–	–	Y	Y	
STOCK APPRECIATING FASTER THAN MARKET	Y	N	–	Y	
PRICE INCREASING ON EXPANDING VOLUME	–	–	Y	Y	
PRICE DECREASING ON CONTRACTING VOLUME	–	–	Y	Y	
BUY THE STOCK ON A REGULAR BASIS	X	–	X	–	–
LOOK FOR ANOTHER STOCK	–	X	–	–	–
HOLD BACK	–	–	–	–	X
BUY THE STOCK ON MARGIN	–	–	–	X	–
EXIT	X	X	X	X	X

Figure A9.1f Eliminating contradiction.

Now let's look at rule 1. The only time we should buy stock on a regular basis under this rule is if there's something wrong with the special situation—otherwise, we'd be under rule 6. So let's expand rule 1 to indicate this fact, as is done in Figure A9.1g.

Now, rearranging the conditions, actions, and rules and renumbering the rules in the decision table in Figure A9.1g gives us the decision table in Figure A9.1a.

	1			2	3	6	ELSE
	A	B	C				
MARKET APPRECIATING	Y	Y	Y	Y	N	Y	
SPECIAL SITUATION	Y	Y	N	–	Y	Y	
STOCK APPRECIATING FASTER THAN MARKET	Y	Y	Y	N	–	Y	
PRICE INCREASING ON EXPANDING VOLUME	Y	N	–	–	Y	Y	
PRICE DECREASING ON CONTRACTING VOLUME	N	–	–	–	Y	Y	
BUY THE STOCK ON A REGULAR BASIS	X	X	X	–	X	–	–
LOOK FOR ANOTHER STOCK	–	–	–	X	–	–	–
HOLD BACK	–	–	–	–	–	–	X
BUY THE STOCK ON MARGIN	–	–	–	–	–	X	–
EXIT	X	X	X	X	X	X	X

Figure A9.1g Expanding rule 1.

2. Our solution is shown in Figure A9.2.

MAIN	1
OPEN OLD-BALANCE-FILE	X
OPEN ADDITION-FILE	X
OPEN CHANGE-FILE	X
OPEN NEW-BALANCE-FILE	X
OPEN ORDER-FILE	X
OPEN TRANSFER-FILE	X
OPEN ERROR-FILE	X
PERFORM READ-OLD-BALANCE-FILE	X
PERFORM READ-ADDITION-FILE	X
PERFORM READ-CHANGE-FILE	X
PERFORM PROCESSING	X
PERFORM CLEANUP-ADDITIONS	X
CLOSE NEW-BALANCE-FILE	X
CLOSE ORDER-FILE	X
CLOSE TRANSFER-FILE	X
PERFORM CLEANUP-CHANGES	X
CLOSE ERROR-FILE	X
EXIT	X

Figure A9.2 Solution.

PROCESSING	1 2
M SET	Y N
PERFORM GET-NEW-BALANCE-RECORD	– X
PERFORM UPDATE	– X
GO AGAIN	– X
EXIT	X –

GET-NEW-BALANCE-RECORD	1 2 3 4
A SET	Y N N N
OLD-BALANCE-PART-NUMBER < ADDITION-PART-NUMBER	– Y N N
OLD-BALANCE-PART-NUMBER = ADDITION-PART-NUMBER	– – Y N
MOVE ADDITION-PART-NUMBER TO ERROR-PART-NUMBER	– – X –
MOVE 'ONE' TO ERROR-KEY	– – X –
WRITE ERROR-RECORD	– – X –
PERFORM READ-ADDITION-FILE	– – X –
GO AGAIN	– – X –
PERFORM PREPARE-ADDITION	– – – X
MOVE OLD-BALANCE-RECORD TO NEW-BALANCE-RECORD	X X – –
PERFORM READ-OLD-BALANCE-FILE	X X – –
EXIT	X X – X

PREPARE-ADDITION	1
MOVE ADDITION-RECORD TO NEW-BALANCE-RECORD	X
PERFORM READ-ADDITION-FILE	X
EXIT	X

Figure A9.2 Solution (continued), page 2.

UPDATE	1 2 3 4 5 6 7 8
C SET	Y N N N N N N N
NEW-BALANCE-PART-NUMBER < CHANGE-PART-NUMBER	– Y N N N N N N
NEW-BALANCE-PART-NUMBER = CHANGE-PART-NUMBER	– Y Y Y Y Y Y N
RECEIPT	– – Y N N N N –
ISSUE	– – – Y N N N –
PART NUMBER CHANGE	– – – – Y N N –
DELETION	– – – – – Y N –
ADD RECEIPT-QUANTITY TO ON-HAND-QUANTITY	– – X – – – – –
SUBTRACT RECEIPT-QUANTITY FROM ON-ORDER-QUANTITY	– – X – – – – –
SUBTRACT ISSUE-QUANTITY FROM ON-HAND-QUANTITY	– – – X – – – –
SUBTRACT ISSUE-QUANTITY FROM AVAILABLE-QUANTITY	– – – X – – – –
ADD ISSUE-QUANTITY TO USAGE-THIS-YEAR	– – – X – – – –
MOVE CHANGE-PART-NUMBER TO ERROR-PART-NUMBER	– – – – – – X X
MOVE 'TWO' TO ERROR-KEY	– – – – – – X –
MOVE 'THREE' TO ERROR-KEY	– – – – – – – X
WRITE ERROR-RECORD	– – – – – – X X
CHANGE NEW-BALANCE-PART-NUMBER	– – – – X – – –
PERFORM ORDER-PROCESSING	X X – – X – – –
WRITE NEW-BALANCE-RECORD ON TRANSFER-FILE	– – – – X – – –
PERFORM READ-CHANGE-FILE	– – X X X X X X
GO AGAIN	– – X X – – X X
WRITE NEW-BALANCE-RECORD ON NEW-BALANCE-FILE	X X – – – – – –
EXIT	X X – – X X – –

ORDER-PROCESSING	1 2
AVAILABLE-QUANTITY ≤ REORDER-POINT	Y N
MOVE NEW-BALANCE-PART-NUMBER TO ORDER-PART-NUMBER	X –
MOVE NEW-BALANCE-DESCRIPTION TO ORDER-DESCRIPTION	X –
MOVE REORDER-POINT TO QUANTITY-ORDERED	X –
ADD REORDER-QUANTITY TO QUANTITY-ORDERED	X –
SUBTRACT AVAILABLE-QUANTITY FROM QUANTITY-ORDERED	X –
WRITE ORDER-RECORD	X –
ADD QUANTITY-ORDERED TO ON-ORDER QUANTITY	X –
ADD QUANTITY-ORDERED TO AVAILABLE-QUANTITY	X –
EXIT	X X

Figure A9.2 Solution (continued), page 3.

CLEANUP-ADDITIONS	1 2
A SET	Y N
PERFORM PREPARE-ADDITION	– X
PERFORM UPDATE	– X
GO AGAIN	– X
EXIT	X –

CLEANUP-CHANGES	1 2
C SET	Y N
MOVE CHANGE-PART-NUMBER TO ERROR-PART-NUMBER	– X
MOVE 'THREE' TO ERROR-KEY	– X
WRITE ERROR-RECORD	– X
PERFORM READ-CHANGE-FILE	– X
GO AGAIN	– X
EXIT	X –

READ-OLD-BALANCE-FILE	1 2 3
ANOTHER OLD-BALANCE-RECORD	Y Y N
FIRST OLD-BALANCE-RECORD	Y N –
READ OLD-BALANCE-FILE	X X –
PERFORM SEQUENCE-CHECK-OLD-BALANCE-FILE	– X –
MOVE OLD-BALANCE-PART-NUMBER TO PREVIOUS-OLD-BALANCE-PART-NUMBER	X X –
CLOSE OLD-BALANCE-FILE	– – X
SET M	– – X
EXIT	X X X

SEQUENCE-CHECK-OLD-BALANCE-FILE	1 2
OLD-BALANCE-PART-NUMBER > PREVIOUS-OLD-BALANCE-PART-NUMBER	Y N
MOVE OLD-BALANCE-PART-NUMBER TO ERROR-PART-NUMBER	– X
MOVE FOUR' TO ERROR-KEY	– X
WRITE ERROR-RECORD	– X
GO TO READ-OLD-BALANCE-FILE	– X
EXIT	X –

Figure A9.2 Solution (continued), page 4.

READ-ADDITION-FILE	1	2	3
ANOTHER ADDITION-RECORD	Y	Y	N
FIRST ADDITION-RECORD	Y	N	–
READ ADDITION-FILE	X	X	–
PERFORM SEQUENCE-CHECK-ADDITION-FILE	–	X	–
MOVE ADDITION-PART-NUMBER TO PREVIOUS-ADDITION-PART-NUMBER	X	X	–
CLOSE ADDITION-FILE	–	–	X
SET A	–	–	X
EXIT	X	X	X

SEQUENCE-CHECK-ADDITION-FILE	1	2
ADDITION-PART-NUMBER $>$ PREVIOUS-ADDITION-PART-NUMBER	Y	N
MOVE ADDITION-PART-NUMBER TO ERROR-PART-NUMBER	–	X
MOVE 'FIVE' TO ERROR-KEY	–	X
WRITE ERROR-RECORD	–	X
GO TO READ-ADDITION-FILE	–	X
EXIT	X	–

READ-CHANGE-FILE	1	2	3
ANOTHER CHANGE-RECORD	Y	Y	N
FIRST CHANGE-RECORD	Y	N	–
READ CHANGE-FILE	X	X	–
PERFORM SEQUENCE-CHECK-CHANGE-FILE	–	X	–
MOVE CHANGE-PART-NUMBER TO PREVIOUS-CHANGE-PART-NUMBER	X	X	–
CLOSE CHANGE-FILE	–	–	X
SET C	–	–	X
EXIT	X	X	X

SEQUENCE-CHECK-CHANGE-FILE	1	2
CHANGE -PART-NUMBER $\geqslant$ PREVIOUS-CHANGE-PART-NUMBER	Y	N
MOVE CHANGE-PART-NUMBER TO ERROR-PART-NUMBER	–	X
MOVE 'SIX' TO ERROR-KEY	–	X
WRITE ERROR-RECORD	–	X
GO TO READ-CHANGE-FILE	–	X
EXIT	X	–

Figure A9.2 Solution (continued), page 5.

Chapter 10

1. In an intimate relationship the relationship is more important than any goals to be attained through the relationship. In a practical relationship the goal is more important than the relationship.

2. For a practical relationship to be successful, the parties must appear credible in each other's eyes.

3. Given your credibility, to get the other person to productively enter a practical relationship you must appeal to his self esteem.

4. Three ways of making another person feel important are:

a. Ask him to do a favor for you in an area that interests him and about which he knows something.

b. Be polite.

c. Use his name when you speak with him.

5. Three things you shouldn't do are:

a. Tell jokes that make the other person look ridiculous.

b. Expose motives of which the other person wouldn't be proud.

c. Condemn him for doing bad things.

6. If you want another person's cooperation in attaining a goal, you can get him interested by showing him what he stands to gain, from his point of view, by cooperating.

7. Five ways of strengthening a relationship with another person are:

a. Be pleasant and cheerful.

b. Avoid topics to which he's sensitive.

c. Do small favors for him.

d. Physical contact.

e. Let him do the talking.

8. When someone brings a complaint to you, hear him out with sympathy and respect.

9. A three-step procedure for avoiding arguments is as follows.

a. First, if you can concede the point, do so.

b. If you can't, try to postpone the question.

c. If you can't, appeal to the facts.

1) Don't tell the other person he's wrong.

2) Try to find out why he feels as he does.

10. A person has bought your idea when he stops discussing it and starts talking about the details of implementing it.

11. When someone criticizes you, take the blame.

12. A three-step procedure for repairing a relationship damaged by your inadequacy is as follows.

a. Admit your failure, apologize, and make an appeal to resolve differences.

b. If the other person won't speak to you, appeal through a disinterested party.

c. If even the mention of your name angers the other person, let some time pass.

13. Three things you can do to maintain another person's self-respect when you must frustrate him are:

a. Give him reasons for your actions that don't attack his sense of importance.

b. Show that you respect him personally.

c. Suggest some other ways he can attain his goal.

Chapter 11

1. Here are the ages:

Joyce (J)	8
Debbie (D)	5
Connie (C)	32
Bill (B)	37
Patti (P)	10
Anne (A)	6
Rag doll (R)	3

This exercise can be solved mechanically by use of algebra. The important thing is to get the problem properly defined. Notice that Patti and Anne, as well as Joyce and Debbie, are Connie's daughters. Thus,

$$C = J + D + P + A + 3$$

not:

$$C = J + D + 3$$

as the first sentence of the problem statement implies.

2. This problem is a classic in the demonstration of the need to make assumptions before conclusions can be deduced from the given facts. The solution is as follows.

> A argues: "Suppose I am blue. Then B will see that I am blue. And knowing that C has not laughed, B will also know that C cannot see two blues. Thus if I am blue, B will know that he is white. But he doesn't know (because he has not raised his hand). Hence I am not blue, but white." (The solution assumes that B or C is sufficiently intelligent to reason in the manner indicated. But if this assumption is unjustified, the failure of B or C to raise his hand has no significance, and there is no possible solution. Hence A has nothing to lose and everything to gain by making the assumption.)

3. The argument mistakes correlation for explanation. To maintain that policemen create traffic jams it would be necessary to demonstrate that, under controlled circumstances, introduction of policemen into traffic situations does cause jams that otherwise don't occur. Until then, a variety of hypotheses are possible, one of which is that jams cause policemen to appear on the scene.

Chapter 12

1. The steps in developing a "profitability index" are:

a. Estimate the life of the proposed data processing system.

b. Estimate the operating cost of the present data processing system.

c. Estimate the operating cost of the proposed system.

d. Subtract the operating cost of the present system from the proposed system operating cost. This gives the operating benefit.

e. Estimate the value of all other benefits of the proposed system.

f. Add the value of all other benefits to the operating benefit. This gives the total benefit.

g. Estimate the scrap value of the proposed system.

h. Add the scrap value to the total benefit. This gives the future benefits of the proposed system.

i. Determine the present value of the future benefits of the proposed system.

j. Estimate the cost of developing the proposed system.

k. Divide the present value of the proposed system by the proposed system development cost. This gives the profitability index of the proposed system.

2. The present data processing system consists of those data processing operations that are scheduled to be discontinued when the proposed system is installed.

3. The operating cost of the present data processing system is the cost of operating the present system over the life of the proposed system.

4. The residual value of a data processing system is the amount by which the development cost of a system designed to replace the data processing system is reduced by retention in the replacement system of certain parts of the data processing system.

5. The present value of a future benefit is the amount you'd be willing to pay today to receive the benefits in the future.

6. The required rate of return is the rate you use to determine the present value of a future benefit.

7. A procedure for quantifying the benefit "reduction in accident rate" might be as follows.

a. Determine the categories of cost experienced when an accident occurs. Some categories might be:

1) Salary, together with overhead, paid to the employee while off the job because of the accident

2) The profit not received because of production lost due to the accident

b. Estimate the extent to which a reduction in accident rate might reduce the cost in each of the above determined categories. One possible approach might be to:

1) Determine the cost in each category for the "average" accident.

2) Estimate the number of accidents that the reduction in accident rate represents.

3) Multiply this number by the costs in each category.

c. Determine the amount by which the reduction in accident rate will reduce insurance premiums.

8. The profitability index is developed as follows.

a. The proposed system is expected to be in operation in years 4 through 7.

b. *Present System Operating Cost (thousands)*

Year	4	5	6	7
	$666	733	806	887

c. *Proposed System Operating Cost (thousands)*

Year	4	5	6	7
	$400	440	484	532

d. *Operating Benefit (thousands)*

Year	4	5	6	7
	$666	733	806	887
	400	440	484	532
	$266	293	322	355

e. *Other Benefits (thousands)*

Year	4	5	6	7
	$200	220	242	266

f. *Total Benefits (thousands)*

Year	4	5	6	7
	$200	220	242	266
	266	293	322	355
	$466	513	564	621

g. Residual Value

0.1($500,000) = $50,000

h.

Future Benefits (thousands)					
Year	4	5	6	7	Residual
	$466	513	564	621	50

i.

Present Value of Future Benefits (thousands)

Year		
4	0.572($446) =	$ 255
5	0.497($513) =	$ 255
6	0.432($564) =	$ 244
7	0.376($621) =	$ 233
Residual	0.376($50) =	$ 19
		$1,006

j. Profitability Index

$$\frac{\$1,006,000}{\$500,000} = 2.01$$

9. The system is economically feasible, because its profitability index is greater than 1.

10. The cash flow analysis is shown in Figure A12.10.

YEAR	1	2	3	4	5	6	7
BEGINNING BALANCE	—	(100)	(300)	(500)	(34)	479	1043
RECEIPTS	—	—	—	866	953	1048	1153
DEVELOPMENT COSTS	100	200	200	—	—	—	—
OPERATING COSTS	—	—	—	400	440	484	532
ENDING BALANCE	(100)	(300)	(500)	(34)	479	1043	1664

Figure A12.10 Cash flow analysis in thousands of dollars.

APPENDICES

APPENDIX A
FLOWCHART OF EXERCISE 2
IN CHAPTER 9

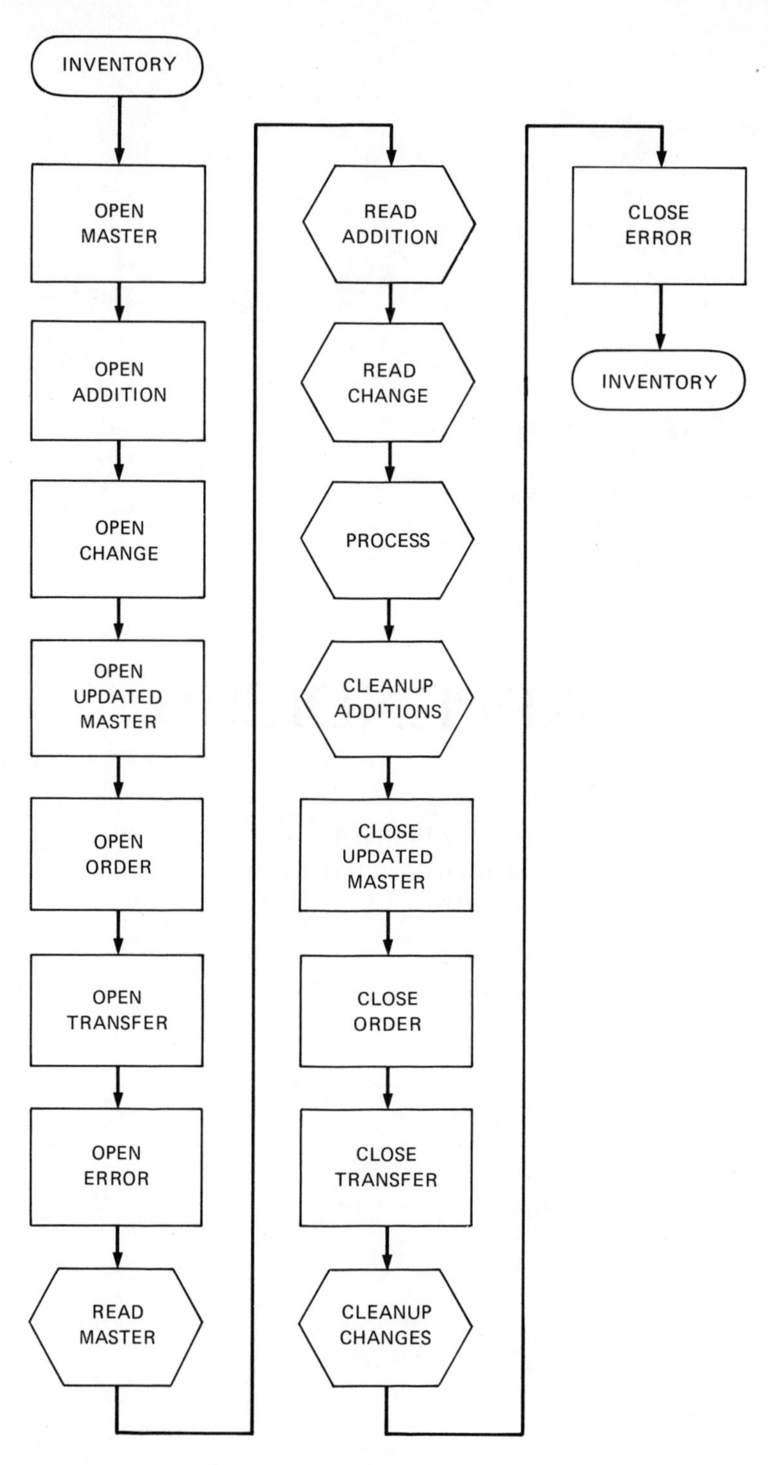
INVENTORY
OPEN MASTER
OPEN ADDITION
OPEN CHANGE
OPEN UPDATED MASTER
OPEN ORDER
OPEN TRANSFER
OPEN ERROR
READ MASTER
READ ADDITION
READ CHANGE
PROCESS
CLEANUP ADDITIONS
CLOSE UPDATED MASTER
CLOSE ORDER
CLOSE TRANSFER
CLEANUP CHANGES
CLOSE ERROR
INVENTORY

Flowchart 1

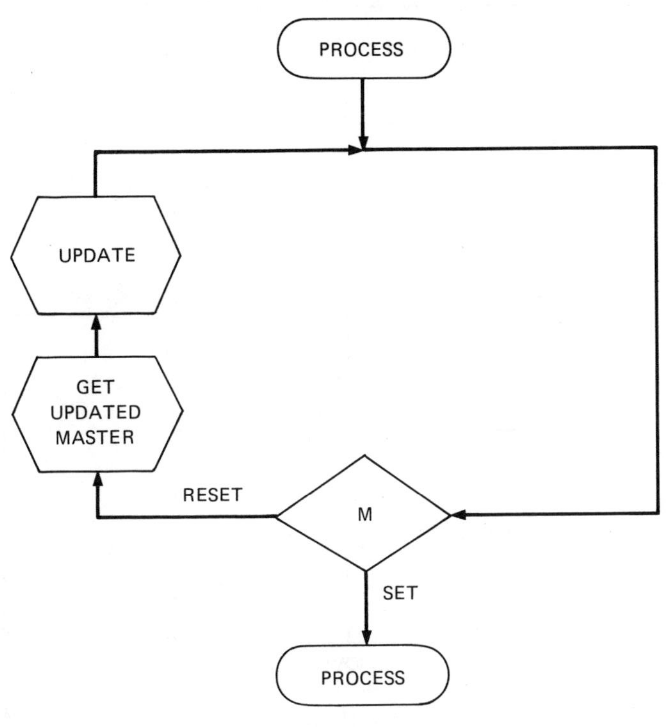

Flowchart 2

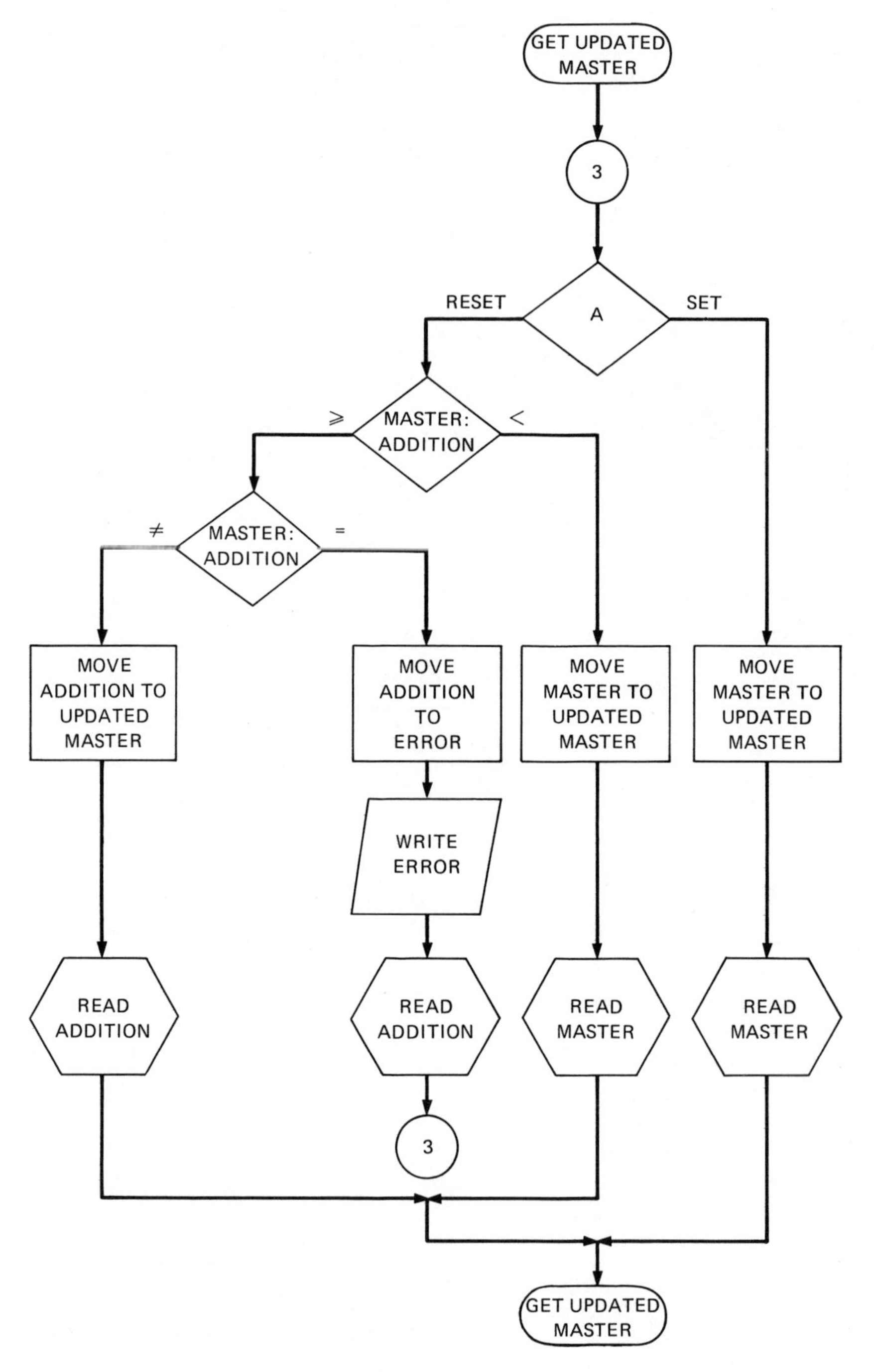
GET UPDATED MASTER
3
RESET
A
SET
≥
MASTER: ADDITION
<
≠
MASTER: ADDITION
=
MOVE ADDITION TO UPDATED MASTER
MOVE ADDITION TO ERROR
MOVE MASTER TO UPDATED MASTER
MOVE MASTER TO UPDATED MASTER
WRITE ERROR
READ ADDITION
READ ADDITION
READ MASTER
READ MASTER
3
GET UPDATED MASTER

Flowchart 3

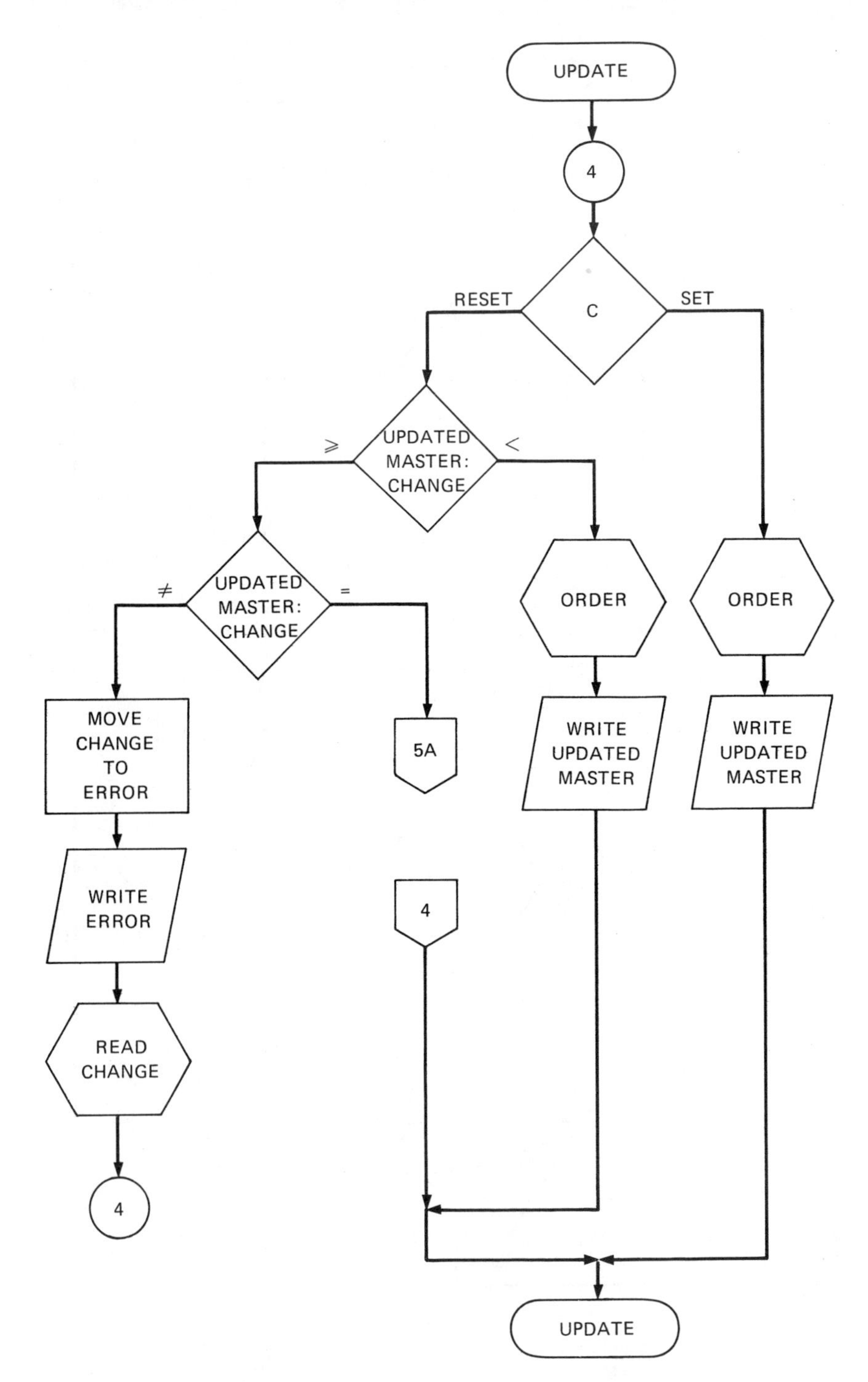
UPDATE
4
RESET
C
SET
UPDATED MASTER: CHANGE
≥
<
UPDATED MASTER: CHANGE
≠
=
ORDER
ORDER
MOVE CHANGE TO ERROR
5A
WRITE UPDATED MASTER
WRITE UPDATED MASTER
WRITE ERROR
4
READ CHANGE
4
UPDATE

Flowchart 4

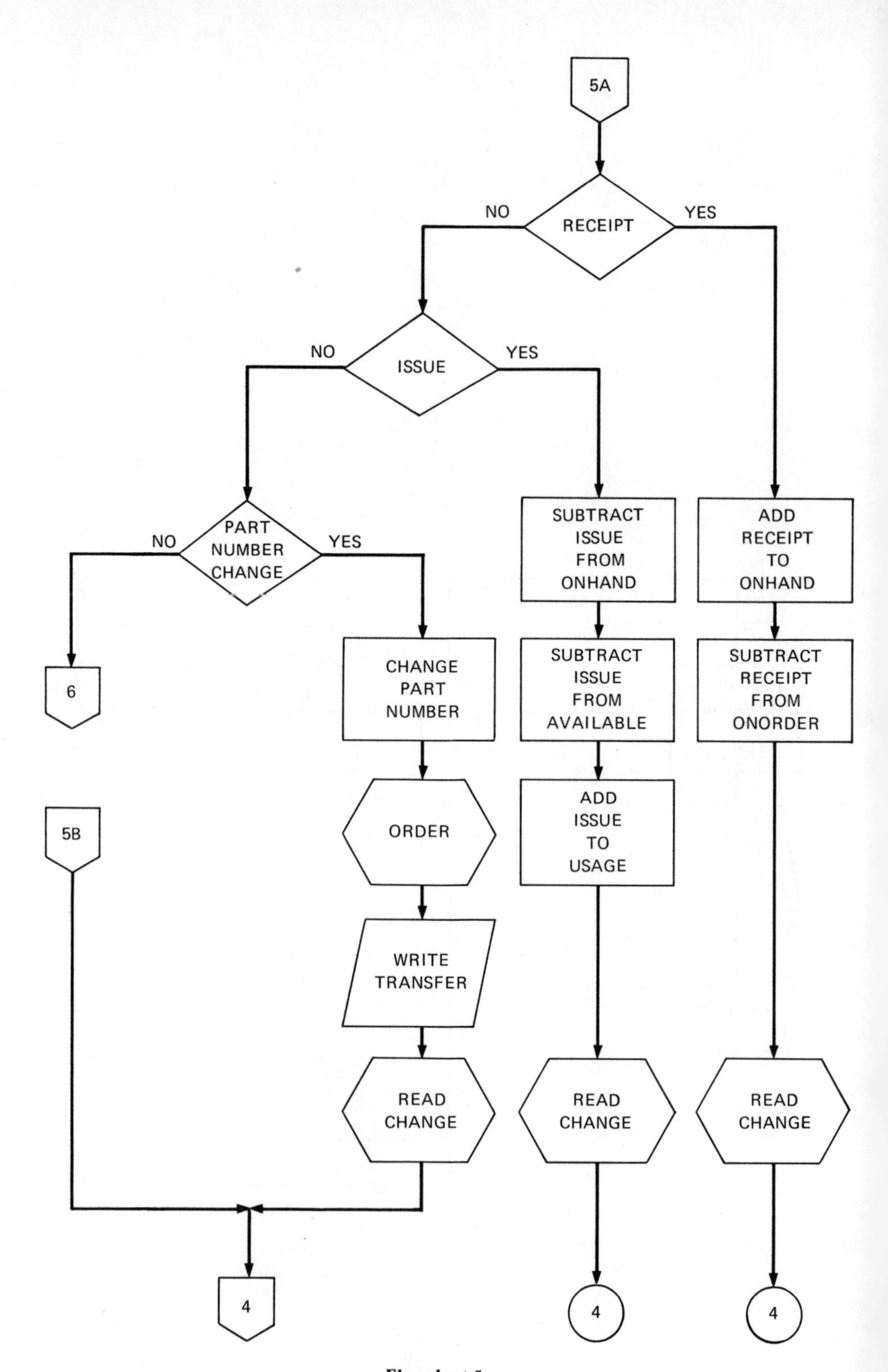
5A
RECEIPT
NO
YES
ISSUE
NO
YES
PART NUMBER CHANGE
NO
YES
SUBTRACT ISSUE FROM ONHAND
ADD RECEIPT TO ONHAND
6
CHANGE PART NUMBER
SUBTRACT ISSUE FROM AVAILABLE
SUBTRACT RECEIPT FROM ONORDER
5B
ORDER
ADD ISSUE TO USAGE
WRITE TRANSFER
READ CHANGE
READ CHANGE
READ CHANGE
4
4
4

Flowchart 5

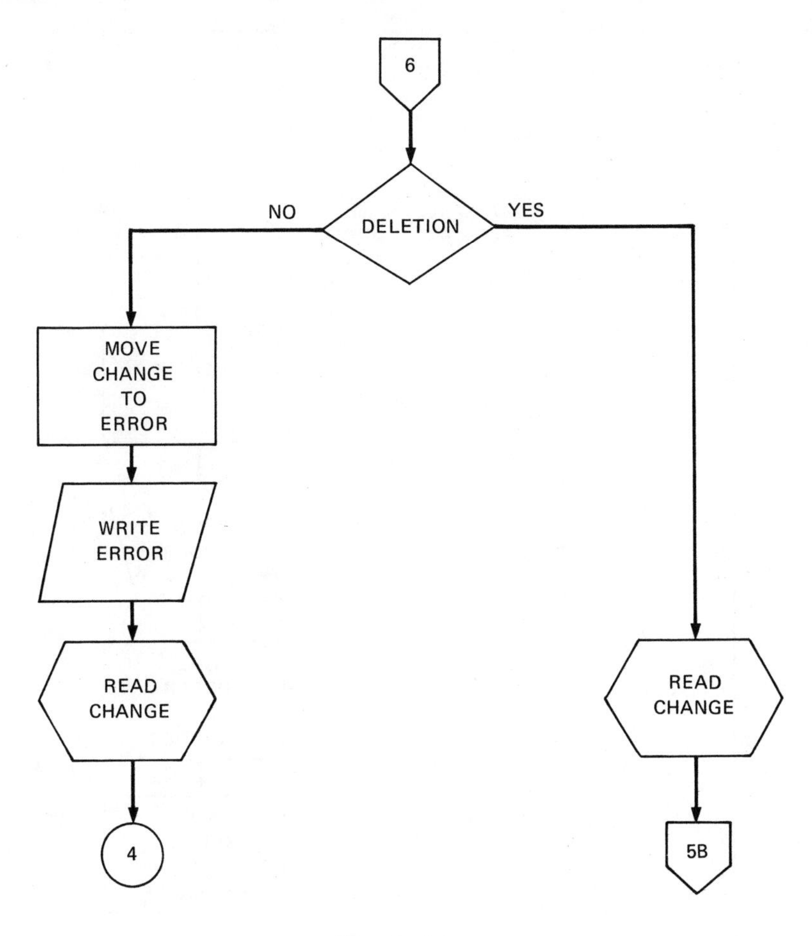
6
DELETION
NO
YES
MOVE
CHANGE
TO
ERROR
WRITE
ERROR
READ
CHANGE
4
READ
CHANGE
5B

Flowchart 6

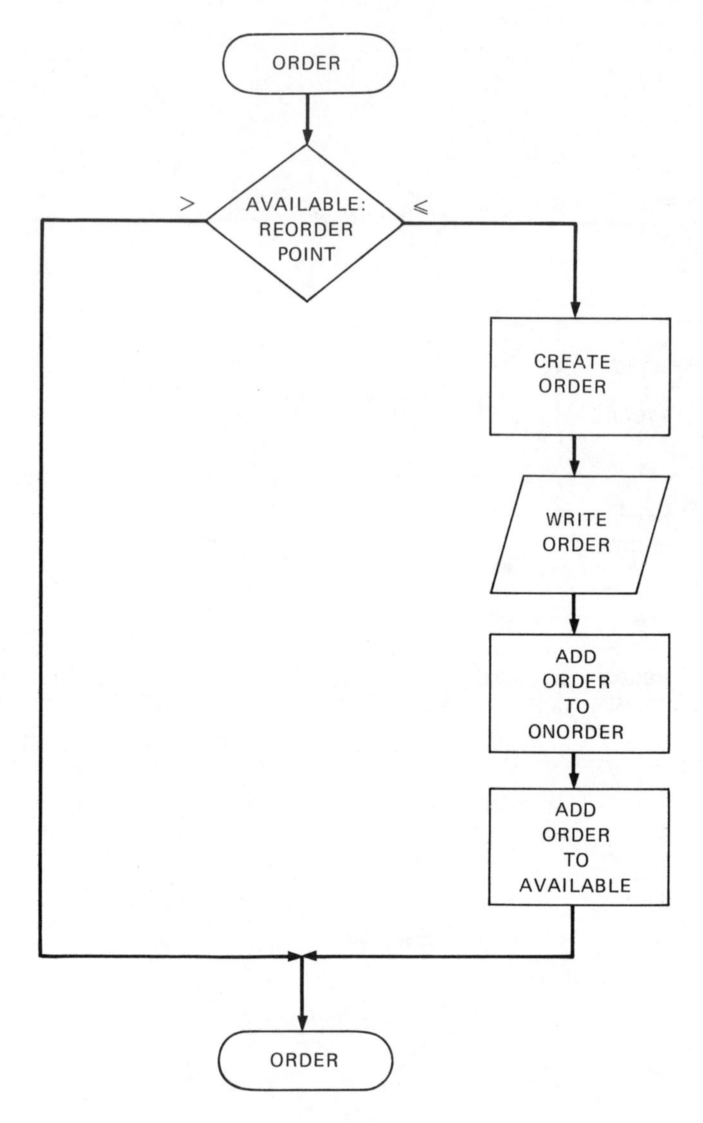
ORDER
AVAILABLE:
REORDER
POINT
>
≤
CREATE
ORDER
WRITE
ORDER
ADD
ORDER
TO
ONORDER
ADD
ORDER
TO
AVAILABLE
ORDER

Flowchart 7

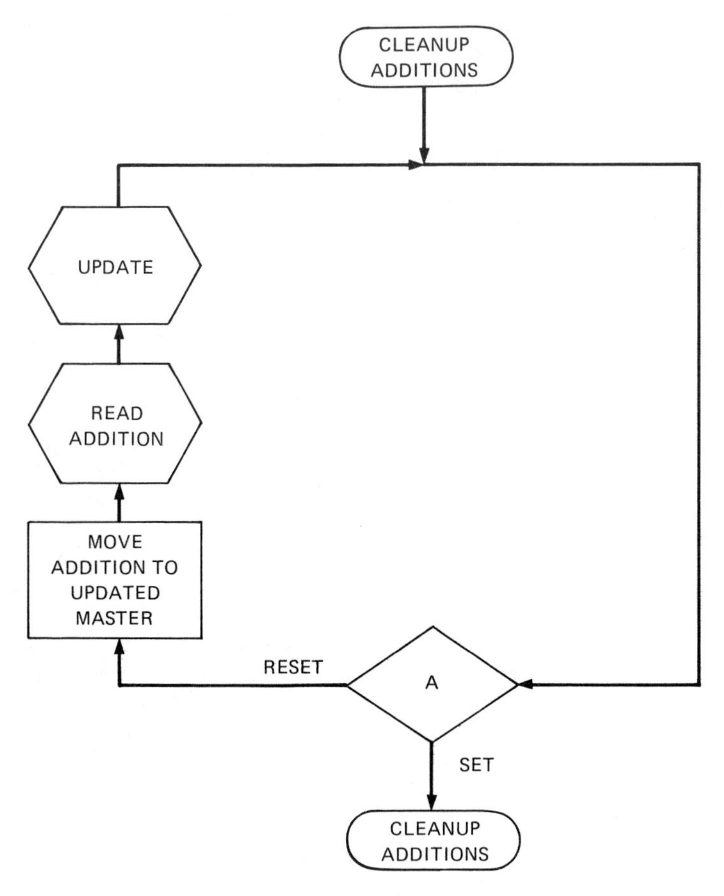
CLEANUP
ADDITIONS
UPDATE
READ
ADDITION
MOVE
ADDITION TO
UPDATED
MASTER
RESET
A
SET
CLEANUP
ADDITIONS

Flowchart 8

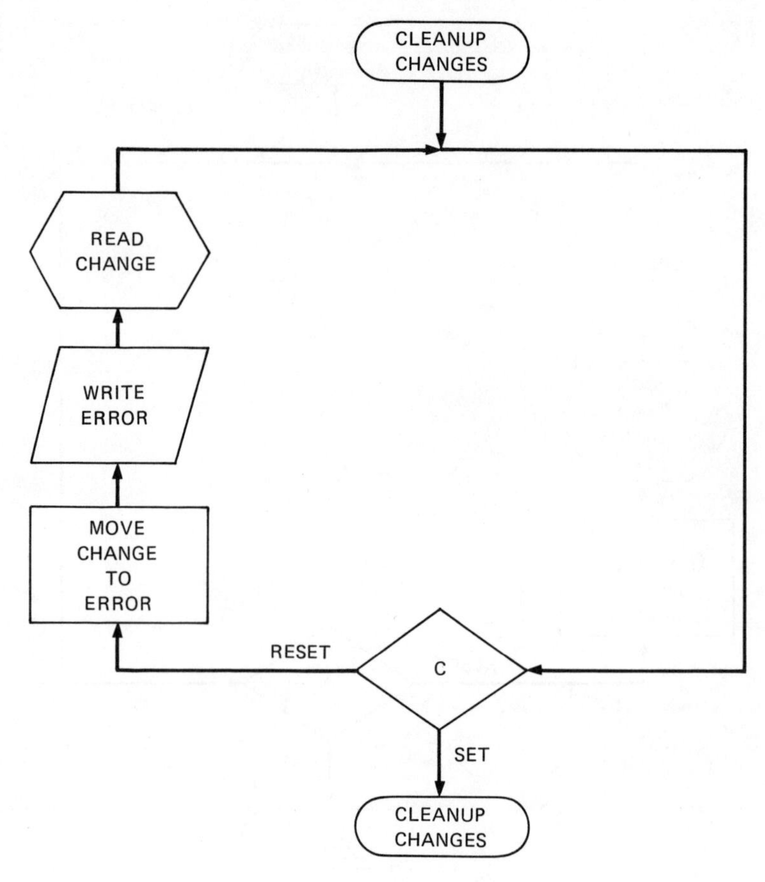
CLEANUP
CHANGES
READ
CHANGE
WRITE
ERROR
MOVE
CHANGE
TO
ERROR
RESET
C
SET
CLEANUP
CHANGES

Flowchart 9

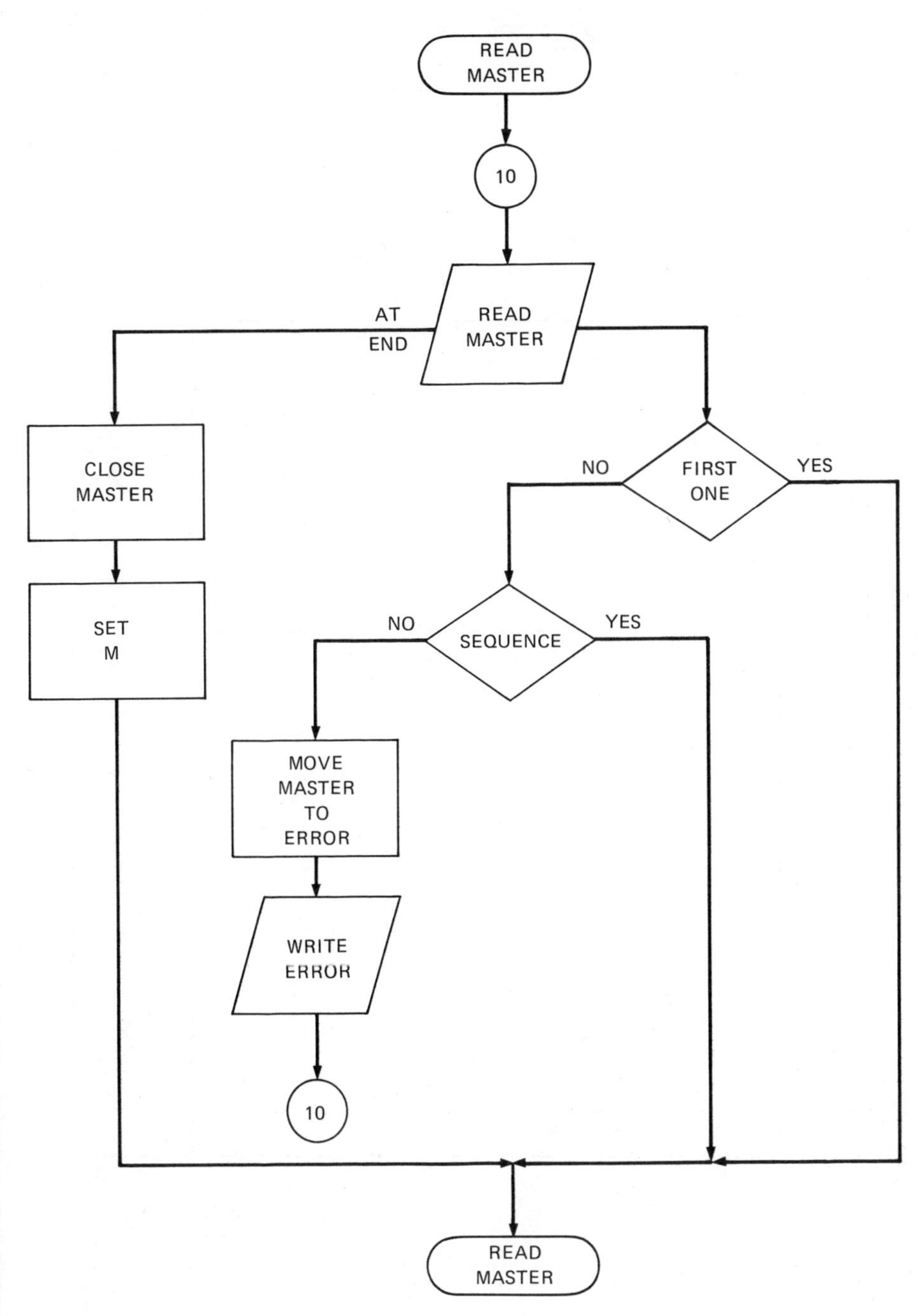

Flowchart 10

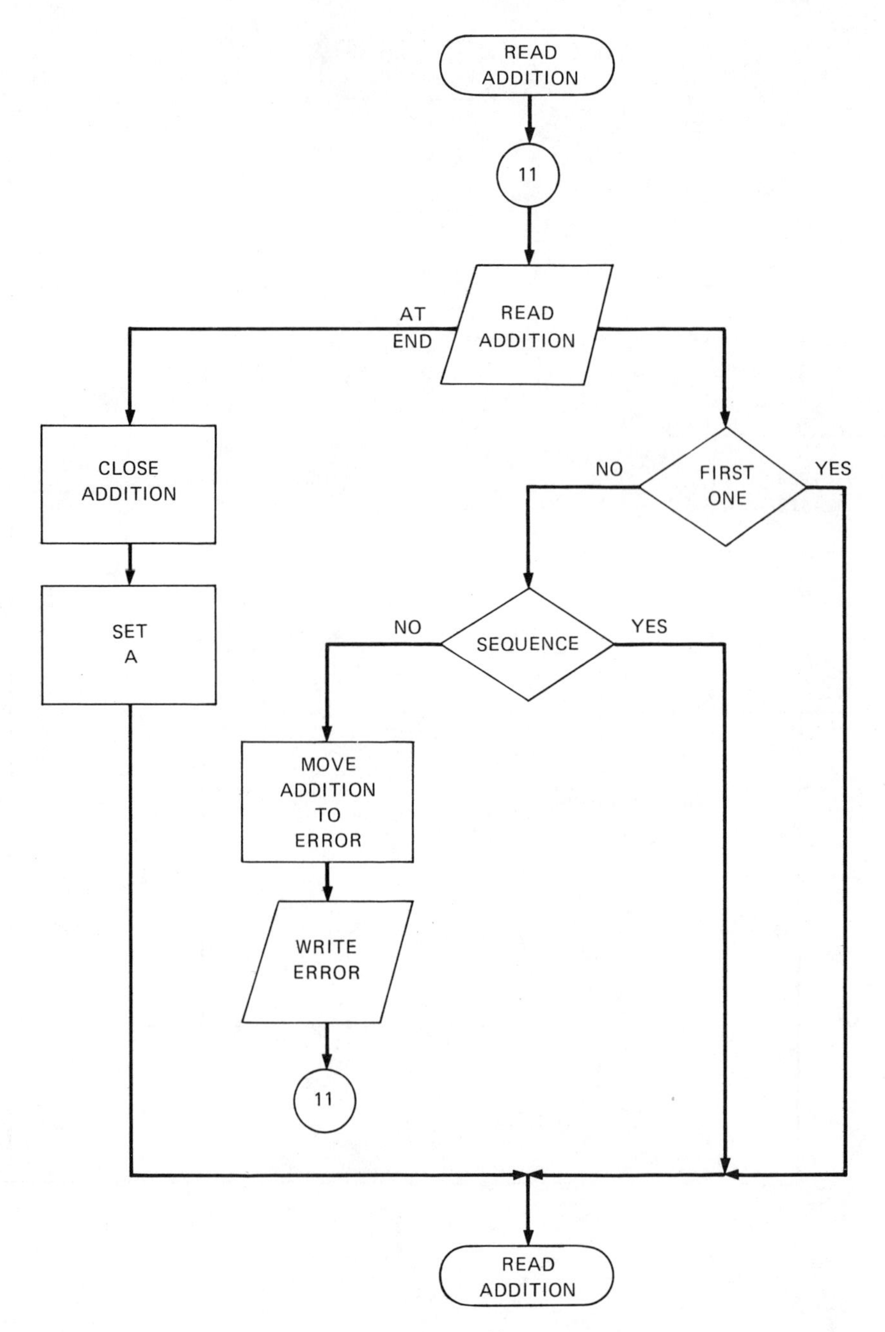

Flowchart 11

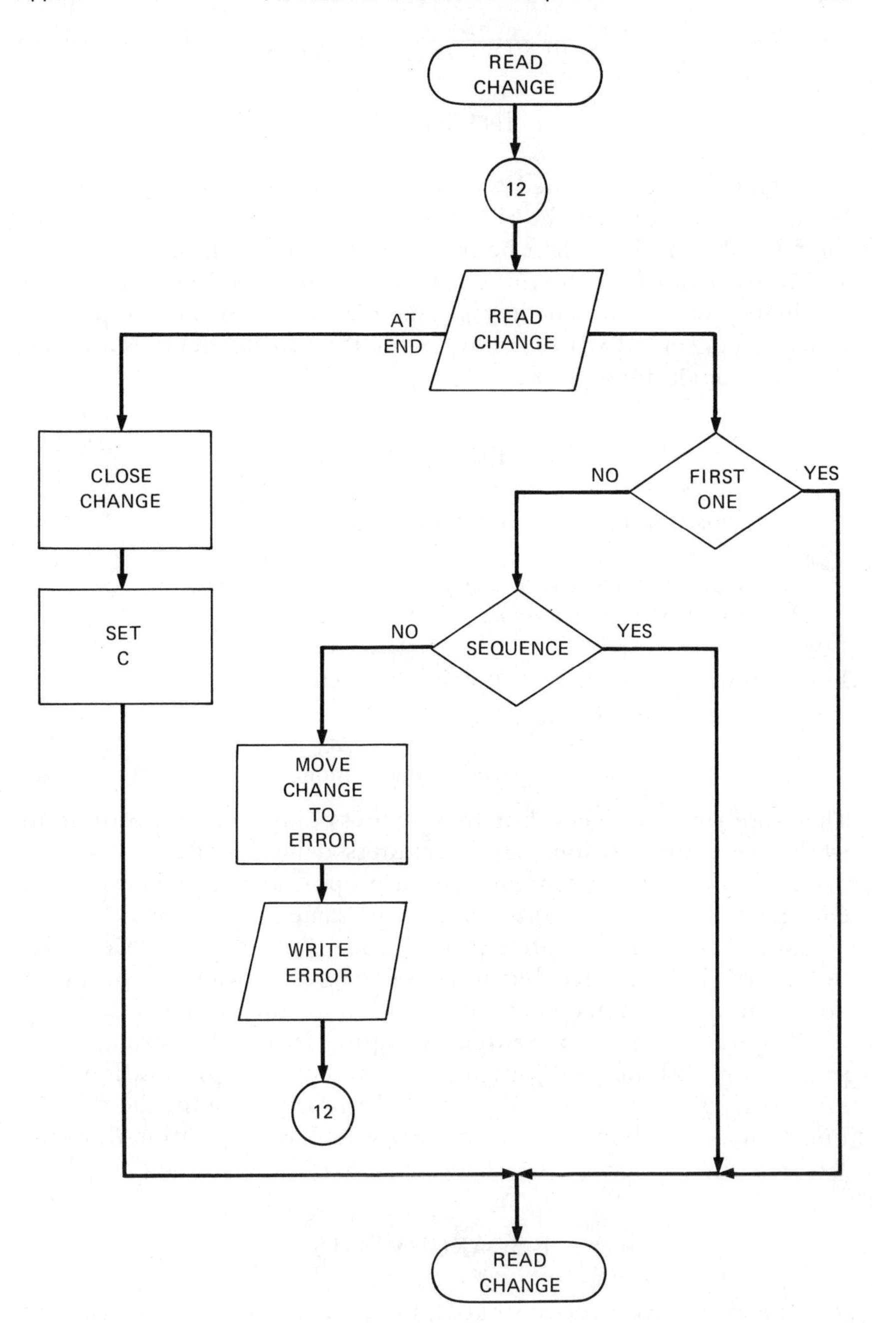

Flowchart 12

APPENDIX B DEDUCTION

B.1. INTRODUCTION

Deduction is concerned with drawing *conclusions* from *premises.* Deduction is useful in the situation where it can be said, "Here are the facts. Now, what might be the consequences of these facts?"

Both *conclusions* and *premises* are *propositions.* Their distinction lies in the way in which they are handled in the deductive process, which is concerned with demonstrating that, if the premises are *true,* then the conclusions are *true.*

B.2. PROPOSITIONS

A *proposition* has two characteristics.

1. It can be expressed in a sentence.
2. It must be either *true* or *false.*

For example, consider the following two sentences.

I'll be paid.

I'll receive remuneration.

These are two sentences, but they express only one proposition. So while every proposition can be expressed in a sentence, there's a distinction between a sentence and a proposition, since many different sentences may be used to express the same proposition.

Not all sentences express propositions. For example, "Code that flowchart" is a sentence, but it doesn't express a proposition, since it doesn't have the characteristic that it's either true or false.

Propositions are represented by capital letters. For example, the proposition "I'll be paid" might be represented as proposition A, or more briefly, as A. To distinguish between propositions, the proposition "I'll be satisfied" might be represented as proposition B, or just B.

B.3. ARGUMENTS

The deductive process that draws conclusions from premises is called an *argument.* A *valid* argument is one which guarantees that if the premises are *true,* the conclusions are *true.* A valid argument has no bearing on the *truth* or *falseness* of the premises.

B.4. TRUTH AND FALSENESS

A proposition is *true* if it conforms to the facts. But this definition is just an evasion of the issue. What truth is is a formidable question with which we will not deal here.

Falseness is the opposite of truth.

B.5. VENN DIAGRAMS

A proposition is either true or false. If it's not true, it's false, and vice versa. This situation can be shown in a diagram, as in Figure B.1. In this diagram, the region inside the box represents the universe of possibilities. The region inside the oval represents the situations in which A is true, and the region inside the box but outside the oval represents the situations in which A is false.

These diagrams are called *Venn diagrams* in honor of the English logician Venn, who first used them.

B.6. NEGATION

Every proposition has a *negation.* The *negation* of a proposition has the opposite value of the proposition. Thus if proposition A is true, then the negation of proposition A is false, and vice versa.

The negation of a proposition is represented by the letter (representing the proposition) followed by a prime. For example, the negation of proposition A is represented as A'.

The symbol A' is read "not A." Thus B' is read "not B."

The relation between a proposition and its negation can be diagramed as shown in Figure B.2. Thus, when A is true, A' is false, and when A is false, A' is true. The whole situation can be summarized in the diagram in Figure B.3.

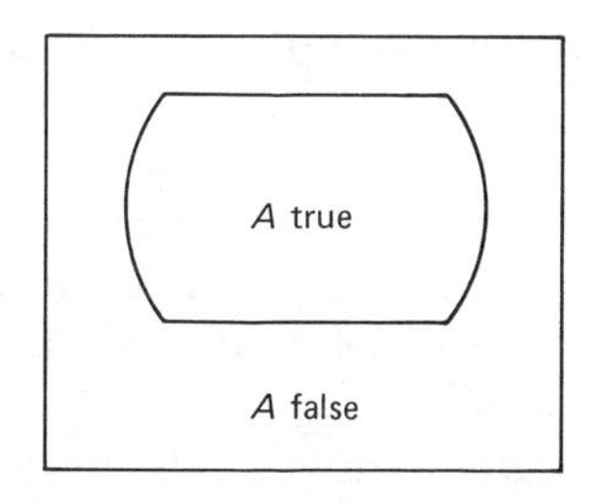

Figure B.1

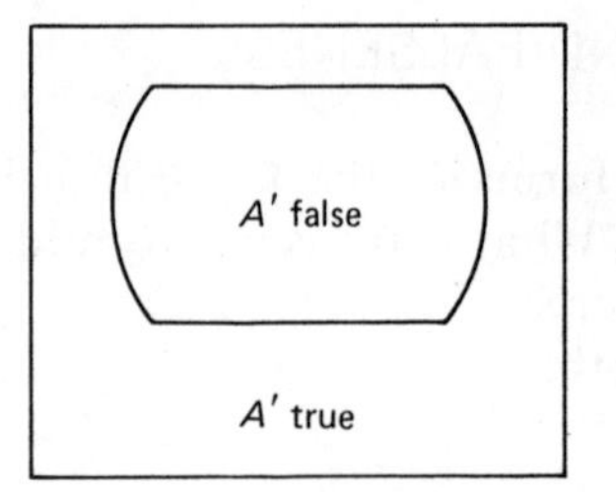

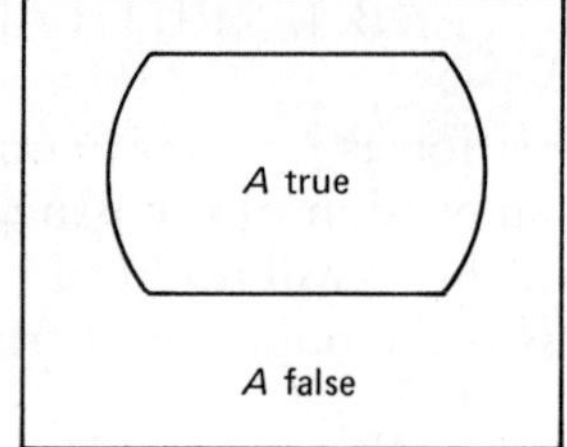

Figure B.2

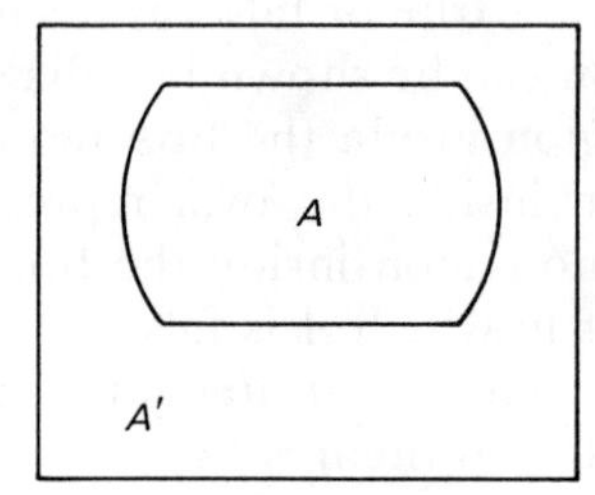

Figure B.3

B.7. CONJUNCTION

Simple propositions can be combined into more complex propositions. For example, given propositions A and B, it may be possible to form the proposition "A and B." Thus, given the propositions:

I'll be paid.
I'll be satisfied.

it's possible to form the proposition "I'll be paid, and I'll be satisfied."

The complex proposition "A and B" is called a *conjunction.* A conjunction is represented by an ampersand connecting the propositions that make it up. For example, the conjunction "A and B" is represented as A & B.

With conjunction and negation it's possible to form some rather complex propositions, such as A & B' and $(A$ & $B')'$. Thus if A is "I'll be paid" and B is "I'll be satisfied," then A & B' is "I'll be paid, and I won't be satisfied," and $(A$ & $B')'$ is "It isn't the case that I'll be paid and I won't be satisfied."

A more complex diagram showing the possibilities for more than one proposition can be constructed as shown in Figure B.4. This diagram isolates four mutually exclusive regions labeled A & B', A' & B, A & B, and A' & B'. The region A & B' represents the situations in which A is true and B is false, the region A' & B the situations in which A is false and B is true, the region A & B the situations in which A and B are both true, and the region A' & B' the situations in which A and B are both false.

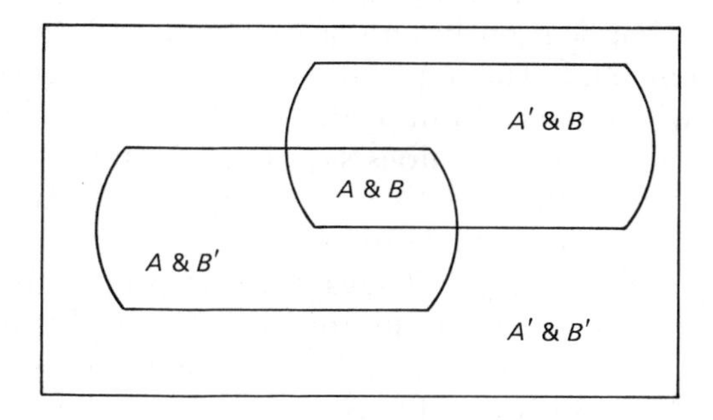

Figure B.4

B.8. IMPLICATION

Another complex proposition is "If A, then B." The proposition "If A, then B" is called an *implication.*

The first proposition in an implication is called the *antecedent.* For example, in the implication "If A, then B," the proposition A is the antecedent.

The second proposition in an implication is called the *consequent.* For example, in the implication "If A, then B," the proposition B is the consequent.

An implication is represented by an arrow connecting the antecedent to the consequent. For example, the implication "If A, then B" is represented as $A \rightarrow B$.

Given the implication $A \rightarrow B$, what can be said concerning the propositions A and B? The possibilities are:

1. A & B
2. A & B'
3. A' & B
4. A' & B'

To take an example, if A is "I'll be paid" and B is "I'll be satisfied," then $A \rightarrow B$ is "If I'm paid, I'll be satisfied," and the four possibilities are

1. I'll be paid, and I'll be satisfied.
2. I'll be paid, and I won't be satisfied.
3. I won't be paid, and I'll be satisfied.
4. I won't be paid, and I won't be satisfied.

First, the implication says nothing about what will happen if I'm not paid. Thus it has no bearing on possibilities (3) and (4).

Second, the implication has no bearing on possibility (1), since it doesn't say I'll be paid. The implication is a conditional statement. It just says what will happen *if* I'm paid.

However, the implication does say that possibility (2) is false. It says that if I'm paid, I'll be satisfied. Therefore, it can't be the case that I'll be paid and also not be satisfied.

Thus the implication $A \rightarrow B$ says that the proposition A & B' is false. This situation is shown in the diagram in Figure B.5. The hatched region in this diagram indicates the situations that aren't possible given the implication $A \rightarrow B$.

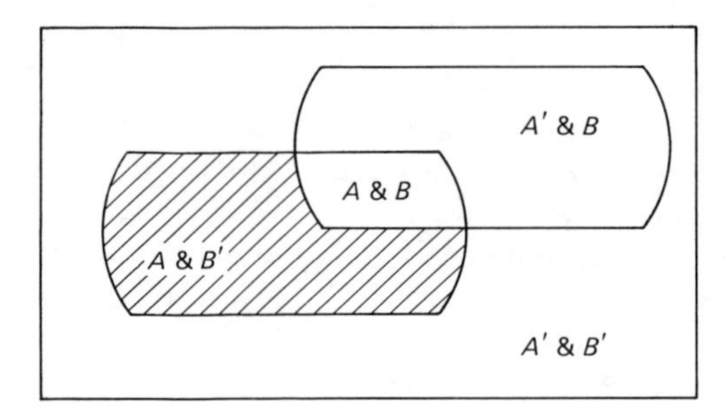

Figure B.5

B.9. ASSERTING THE ANTECEDENT

Now suppose the proposition A (I'll be paid) is true. We can show this in the diagram in Figure B.6. The region with the opposite hatch indicates the situations that are excluded when A is true.

The unhatched region of the diagram indicates that, given the implication $A \rightarrow B$ and given that proposition A is true, then it must be the case that proposition B is true. There are no other possibilities.

This argument can be summarized as follows.

Premise 1: If I'm paid, I'll be satisfied.

Premise 2: I'll be paid.

Conclusion: I'll be satisfied.

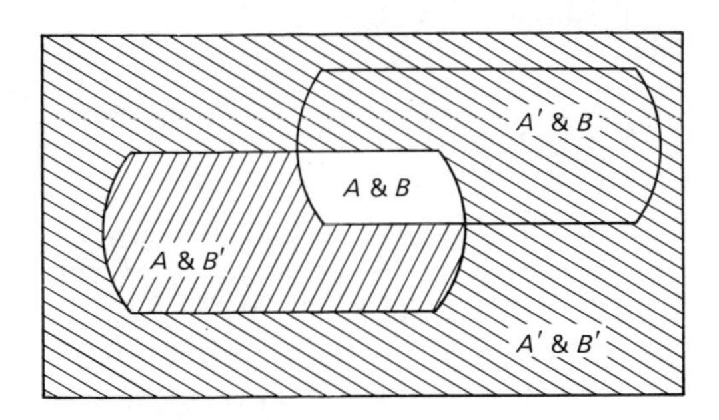

Figure B.6

The form of this argument can be represented as follows.

$$\begin{array}{l} A \rightarrow B \\ \underline{A \qquad} \\ B \end{array}$$

This argument is valid; that is, if the premises are true, the conclusion is true. This form of argument is called *asserting the antecedent.*

B.10. OTHER VALID ARGUMENTS

Now lets investigate some other valid argument forms.

B.10.1. Denying the Consequent

Let's once more take the case where we have the implication $A \rightarrow B$. This situation is shown in Figure B.5.

Now suppose proposition B is false—that is, proposition B' is true. The diagram for this situation is shown in Figure B.7. The region with the opposite hatch indicates the situations that are excluded when B is false.

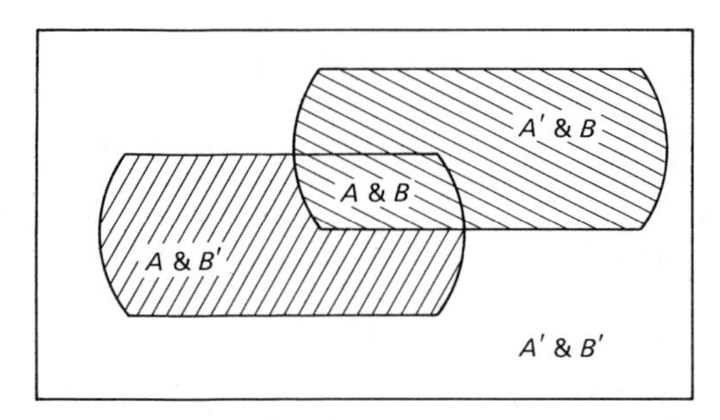

Figure B.7

This diagram indicates that the following argument is valid.

$$\begin{array}{l} A \rightarrow B \\ \underline{B' \qquad} \\ A' \end{array}$$

This form of argument is called *denying the consequent.*

B.10.2. The Chain Argument

A diagram showing the possibilities for three propositions can be constructed as shown in Figure B.8. Suppose we're given the implication $A \rightarrow B$. We know that this excludes the possibility of A & B'. This situation is shown in Figure B.9.

Now suppose we're also given the implication $B \rightarrow C$. This excludes the possibility of B & C', as shown in Figure B.10. This diagram shows that the possibility of A & C' has been excluded. Thus the implication $A \rightarrow C$ is true. Therefore the following is a valid argument.

$$\begin{array}{c} A \rightarrow B \\ B \rightarrow C \\ \hline A \rightarrow C \end{array}$$

This form of argument is called the *chain argument.*

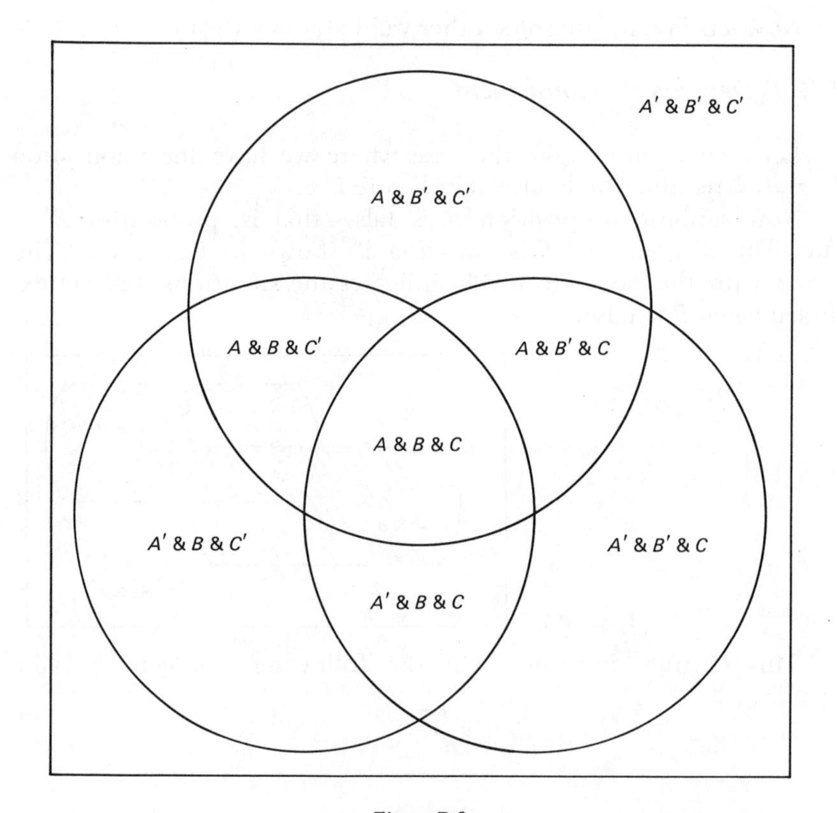

Figure B.8

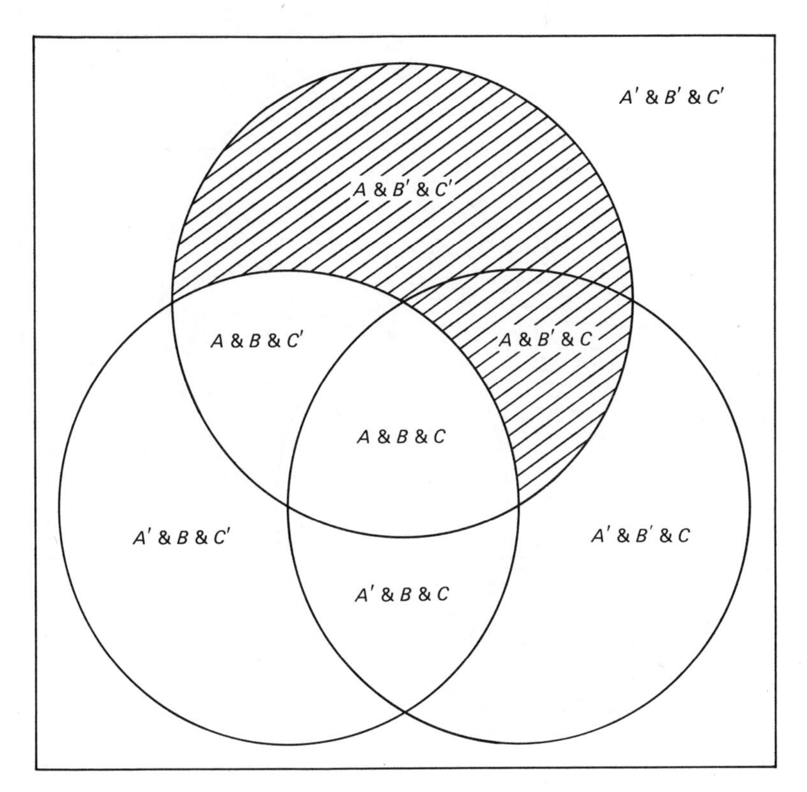

Figure B.9

B.10.3. The Indirect Argument

The *indirect argument* begins by assuming that the proposition you wish to prove isn't true. Thus if you want to prove proposition A, you assume the proposition A'. You must then demonstrate that given A', the proposition A inevitably follows. What you'll then have demonstrated is the implication $A' \rightarrow A$ (that is, "If not-A, then A"). Once you've made this demonstration, it follows that proposition A is true.

Indirect arguments generally involve an argument chain. For example, consider the following.

> So you want to take off Friday. OK, but if you do, you must first complete the work that is due Friday. It's now the end of Thursday, and the work isn't done. Therefore you can't take off Friday.

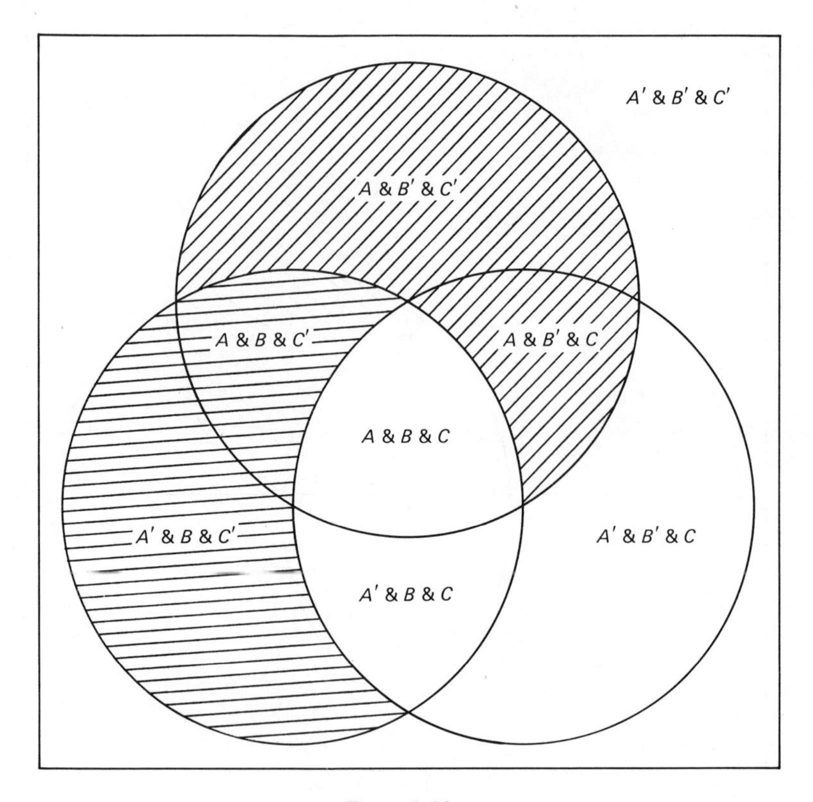

Figure B.10

A more formal statement of this argument would be as follows. Let A be the proposition that you will work on Friday. Then A' is the proposition that you won't work on Friday. Let B be the proposition that you'll get your work done. Then the implication $A' \to B$ is "If you're not going to work on Friday, then you must get your work done," the implication $B \to A$ is "If you're going to get your work done, you'll have to work on Friday," and the argument has the form:

$$\begin{array}{l} A' \to B \\ B \to A \end{array}$$

But this is a chain argument, and consequently a valid conclusion is the implication $A' \to A$. This implication is the form of an indirect argument, and the contention is that it's valid to draw the conclusion A. That is,

$$\frac{A' \to A}{A}$$

The proof that an indirect argument is valid is as follows. Given the implication $A' \rightarrow A$, then the proposition A' & A' is false. But this statement is redundant; it's sufficient to say the proposition A' is false. And this means the proposition A is true.

B.10.4. Reductio ad Absurdum

The argument by *reductio ad absurdum* also begins by assuming that the proposition you wish to prove isn't true. You must then show that from this one assumption, A', it's possible to conclude that some other proposition is both true and false. Once this is done, you've proved that A is true. More formally stated, the argument by *reductio ad absurdum* proceeds as follows.

$$\begin{array}{l} A' \rightarrow B \\ A' \rightarrow B' \\ \hline A \end{array}$$

For example, suppose we want to prove that $\sqrt{2}$ is an irrational number (proposition A). We assume that $\sqrt{2}$ is a rational number (proposition A'). If such is the case, it must be possible to express 2 as a ratio of two whole numbers. (This is the definition of a rational number.) That is,

$$\sqrt{2} = \frac{m}{n}$$

where m and n are whole numbers.

Let's express the ratio m/n in such a way that the numerator and denominator have no common factors. Then we obtain:

$$\frac{m}{n} = \frac{k\ (p)}{k\ (q)}$$

where the ratio p/q is the same as the ratio m/n except that all common factors have been removed, and k is the reciprocal of the product of the common factors.

Substituting, we obtain:

$$\sqrt{2} = \frac{m}{n} = \frac{k\ (p)}{k\ (q)} = \frac{k}{k}\left(\frac{p}{q}\right) = \frac{p}{q}$$

We now have the numbers p and q with no common factors. This is proposition B, and we've just demonstrated the implication $A' \rightarrow B$ (if $\sqrt{2}$ is a rational number, then p and q have no common factors).

Now we must demonstrate the implication $A' \rightarrow B'$ (if $\sqrt{2}$ is a rational number, the p and q have a common factor). We have:

$$\sqrt{2} = \frac{p}{q}$$

If so, then:

$$2 = \frac{p^2}{q^2}$$

or:

$$2q^2 = p^2$$

Now the number $2q^2$ is an even number. (Any number multiplied by 2 is an even number.) Therefore p^2 is an even number (since it's equal to $2q^2$). The number p^2 can be an even number only if the number p is even. (Only the squares of even numbers are even.) Therefore, we must have:

$$p = 2s$$

where s is one half of p. Now let's do some substitution.

$$2q^2 = p^2 = (2s)^2 = 4s^2$$

$$q^2 = 2s^2$$

Now the number $2s^2$ is an even number. Therefore q^2 is an even number. Therefore q is an even number.

We've now demonstrated that both p and q are even numbers. But if they are, they have the factor 2 in common. We've now demonstrated the implication $A' \rightarrow B'$ (if $\sqrt{2}$ is a rational number, then p and q have a common factor.) Having demonstrated both the implication $A' \rightarrow B$, and the implication $A' \rightarrow B'$, we're now in a position to conclude A ($\sqrt{2}$ is an irrational number).

The proof that the argument of *reductio ad absurdum* is valid is as follows. Given the implication $A' \rightarrow B$, then the proposition A' & B' is false (Figure B.11). Given the implication $A' \rightarrow B'$, then the proposition A' & B is false (Figure B.12). The diagram shows that all possibilities other than those including A have been excluded.

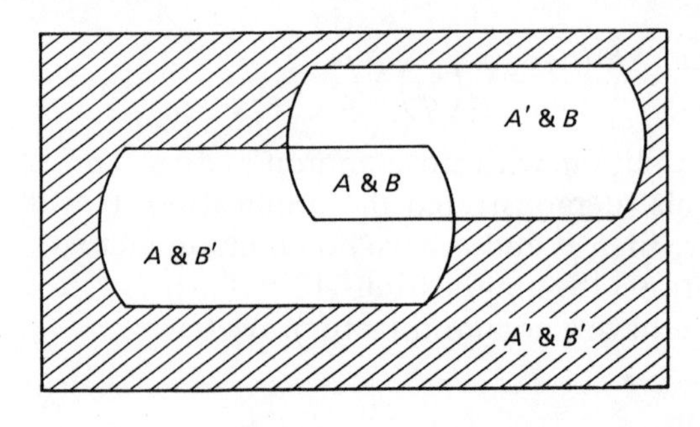

Figure B.11

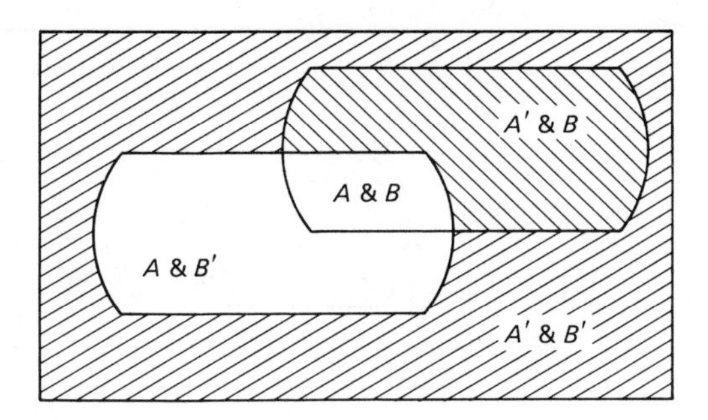

Figure B.12

B.10.5. Dilemma

The form of the *argument by dilemma* is as follows.

$$\begin{array}{l} A \rightarrow B \\ \underline{A' \rightarrow B} \\ B \end{array}$$

The proof of the argument by dilemma is as follows. Given the implication $A \rightarrow B$, then the proposition A & B', is false (Figure B.13). Given the implication $A' \rightarrow B$, then the proposition A' & B' is false (Figure B.14). All remaining possibilities include B.

Like indirect arguments, arguments by dilemma generally involve argument chains. For example:

> If we include this routine in this module, then the module's response time will be unacceptable. If response time is unacceptable, the module is unacceptable. If we don't include the routine in the module, the module won't work. If the module won't work, it's unacceptable. Therefore the module is unacceptable.

Let A be the proposition that the routine is included in the module, B, that the module's response time is unacceptable, C that the module is unacceptable, and D that the module won't work. The argument then has the form:

$$\begin{array}{l} A \rightarrow B \\ B \rightarrow C \\ A' \rightarrow D \\ \underline{D \rightarrow C} \\ C \end{array}$$

The first two implications form a chain that can be reduced to the implication $A \rightarrow C$, the second two a chain that can be reduced to

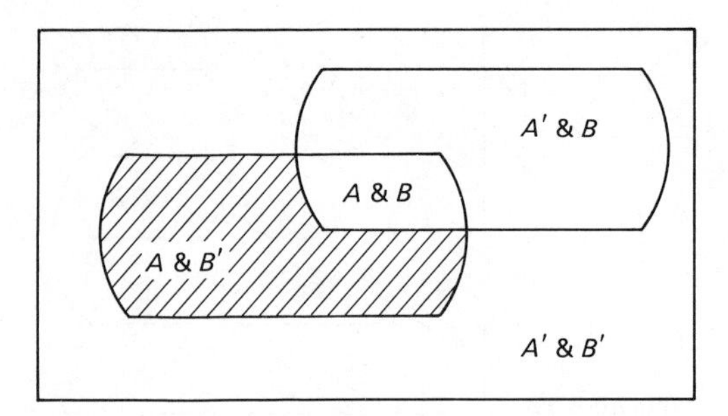

Figure B.13

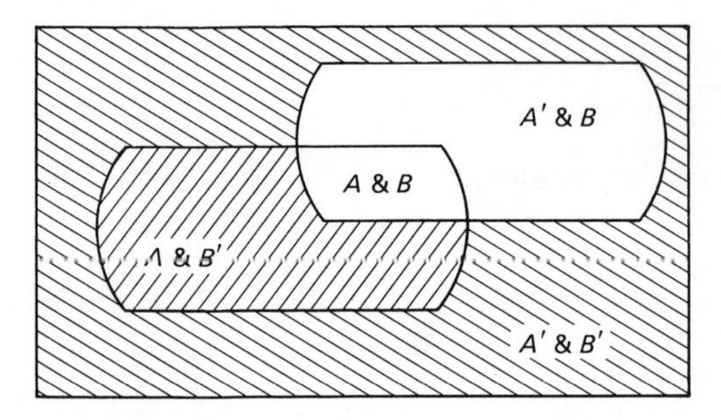

Figure B.14

$A' \rightarrow C$. This returns us to the prototype form of the argument by dilemma.

$$\begin{array}{l} A \rightarrow C \\ \underline{A' \rightarrow C} \\ C \end{array}$$

B.11. ALTERNATION

Before we can continue our investigation of valid argument forms, we must introduce another complex proposition form, the *alternation*. An *alternation* has the form "Either A or B." The "or" here is *exclusive*. That is, it says, "Either A is the case or B is the case, but both of them at once is not the case." Thus an alternation excludes the conjunction A & B as well as the conjunction A' & B' (Figure B.15). Thus, we have either A and not-B, or B and not-A.

An alternation is represented by an inverted carat connecting the propositions that make it up. Thus "Either A or B" is represented as $A \vee B$.

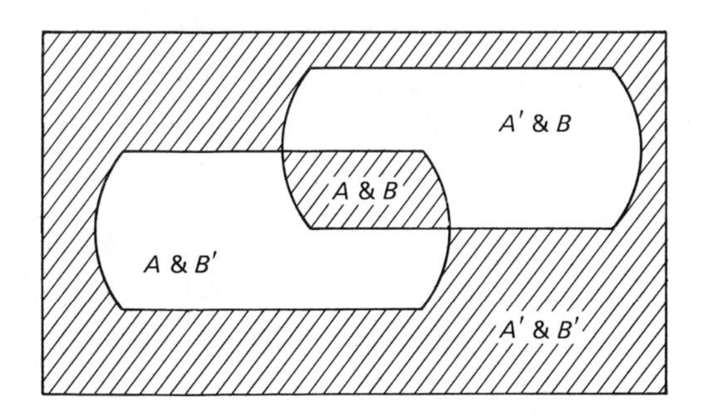

Figure B.15

B.12. THE ALTERNATIVE ARGUMENT

The form of the alternative argument is

$$\frac{\begin{array}{l} A \text{ v } B \\ B' \end{array}}{A}$$

The proof is as follows. Given the alternation A v B, and given that B is not the case, we have the situation diagrammed in Figure B.16. All remaining cases include A.

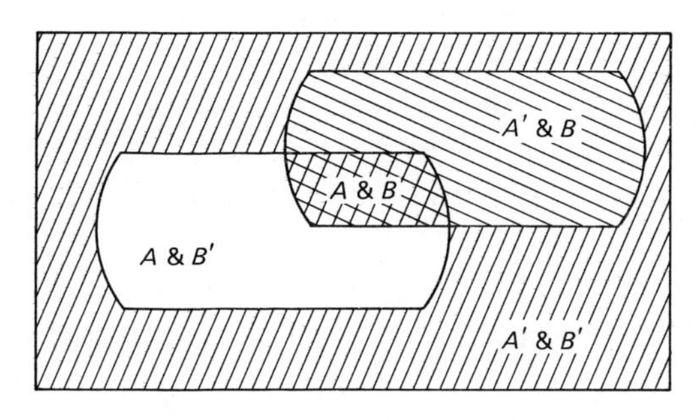

Figure B.16

B.13. INCLUSIVE ALTERNATION

It's worth noting that the alternative argument remains valid even when the alternation in *inclusive,* rather than exclusive. The *inclusive* alternation includes the conjunction A & B as well as the conjunctions A & B' and A' & B. That is, A may be the case, or B may be the case, or both may be the case at once (Figure B.17).

Now, given not-B, we have the situation diagrammed in Figure B.18. And all remaining cases include A.

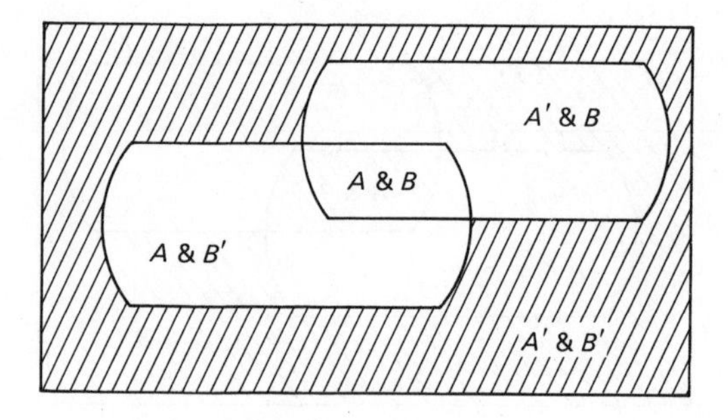

Figure B.17

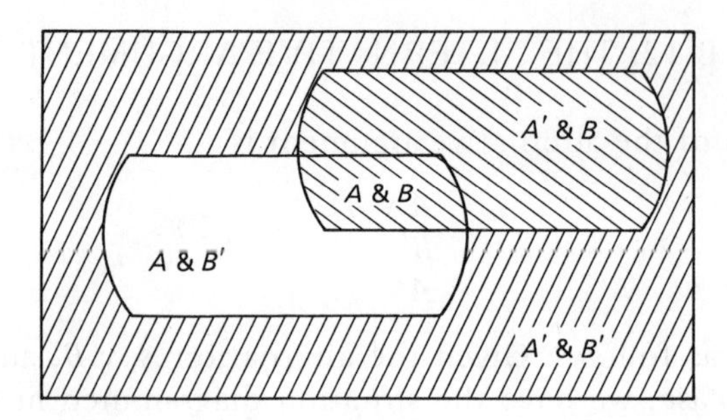

Figure B.18

Thus it's always possible to give an alternation the *weaker* interpretation (the inclusive or is *weaker* than the exclusive or, since it excludes fewer possibilities) and still have a valid alternative argument.

B.14. SUMMARY OF VALID ARGUMENTS

The following is a summary of valid argument forms.
Asserting the antecedent

$$\begin{array}{l} A \rightarrow B \\ \underline{A \qquad} \\ B \end{array}$$

Denying the consequent

$$\begin{array}{l} A \rightarrow B \\ \underline{B' \qquad} \\ A' \end{array}$$

Chain

$$\frac{\begin{array}{l} A \rightarrow B \\ B \rightarrow C \end{array}}{A \rightarrow C}$$

Indirect

$$\frac{A' \rightarrow A}{A}$$

Reductio ad absurdum

$$\frac{\begin{array}{l} A' \rightarrow B \\ A' \rightarrow B' \end{array}}{A}$$

Dilemma

$$\frac{\begin{array}{l} A \rightarrow B \\ A' \rightarrow B \end{array}}{B}$$

Alternative

$$\frac{\begin{array}{l} A \vee B \\ B' \end{array}}{A}$$

B.15. THE LOGIC OF CLASSES

The *logic of classes* is concerned with propositions of the form:

All *a* are *b*

Some *a* are *b*

No *a* are *b*

Some *a* are not *b*

where the lower case letters stand for classes. For example, *a* might be the class of all activities that are to be controlled, and *b* might be all activities that must be planned. Then "All *a* are *b*" would mean "All activities that are to be controlled must be planned."

Propositions of the form "All *a* are *b*" and "No *a* are *b*" are called *universal* propositions. Propositions of the form "Some *a* are *b*" and "Some *a* are not *b*" are called *particular* propositions.

An argument concerning classes takes the form of a *syllogism*. A *syllogism* always takes the form of two premises and a conclusion and always involves three classes. For example,

> Premise: All activities that are to be controlled must be planned.
>
> Premise: Projects must be controlled.
>
> Conclusion: Projects must be planned.

This syllogism involves three classes.

1. Activities that are to be controlled.
2. Activities that must be planned.
3. Projects.

The validity of a syllogism can be determined by using a special kind of Venn diagram, as shown in Figure B.19. Each of the circles represents one of the classes, and the regions into which the figure is divided represent the classes to which an individual case may belong. Thus a case may belong to just one class, to a combination of two classes, or to all three classes. For purposes of reference, these regions might be numbered as shown in Figure B.20.

The Venn diagram is used in the following way to analyze syllogisms involving universal propositions. First the premises are used to hatch out the areas excluded by these premises (Figures B.21 and B.22). The resulting diagram is then inspected to see whether the conclusion follows. In this case the syllogism is valid, since all projects must fall into region 3, which is completely included in the class of activities that must be planned.

Syllogisms involving particular propositions are analyzed in an analogous way except that a bar is used to indicate the areas in which

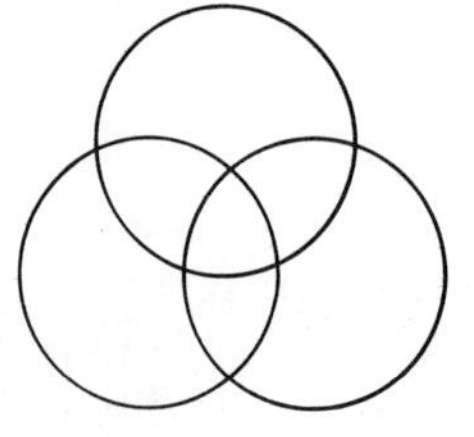

Figure B.19

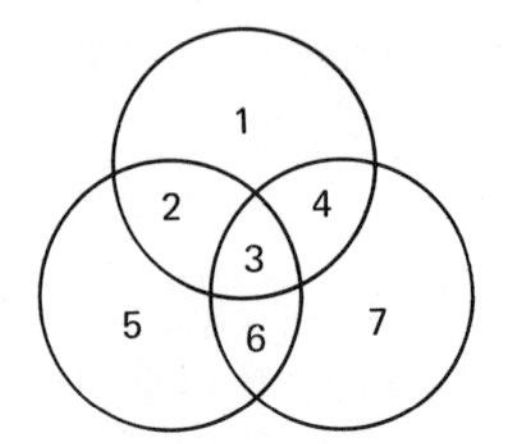

Figure B.20

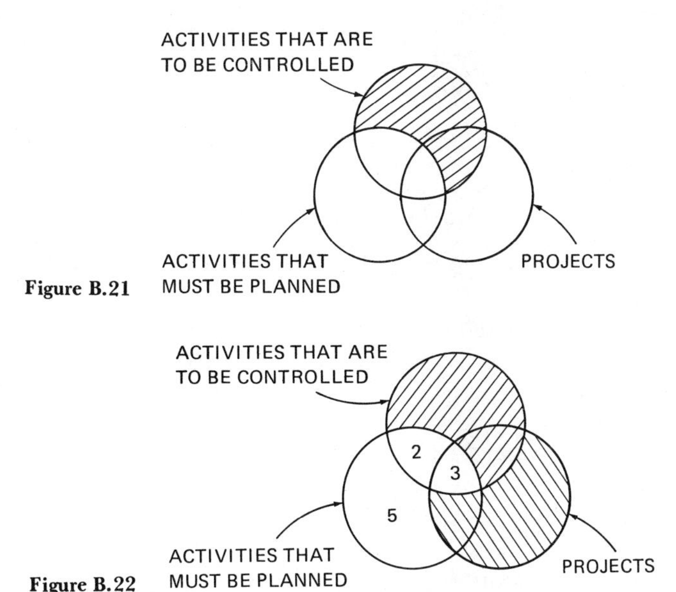

Figure B.21

Figure B.22

a case may appear. Thus the diagram in Figure B.23 indicates that somewhere in the area 3 + 4 there is at least one project, which is all the premise states.

Although it's not necessary, when a syllogism involves one universal premise and one particular premise, it creates a more readable figure to diagram the universal premise before the particular one.

Now for some examples.

1. All *a* are *b*
   All *c* are *b*
   ―――――――
   All *c* are *a*

The conclusion says that all *c* are in area 3 in Figure B.24. But the diagram indicates that this isn't necessarily the case. Some *c* may be in area 6. The argument is invalid.

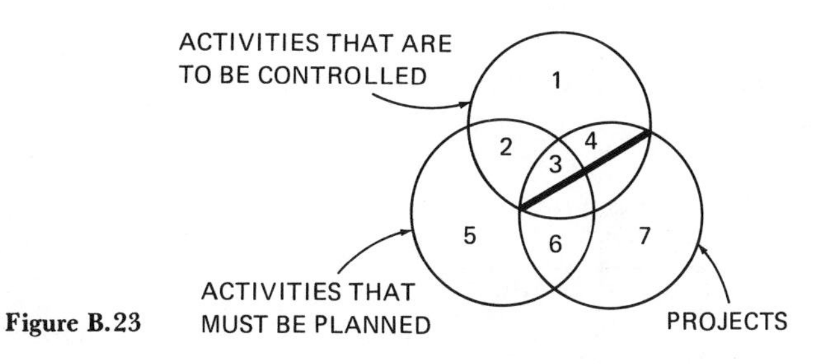

Figure B.23

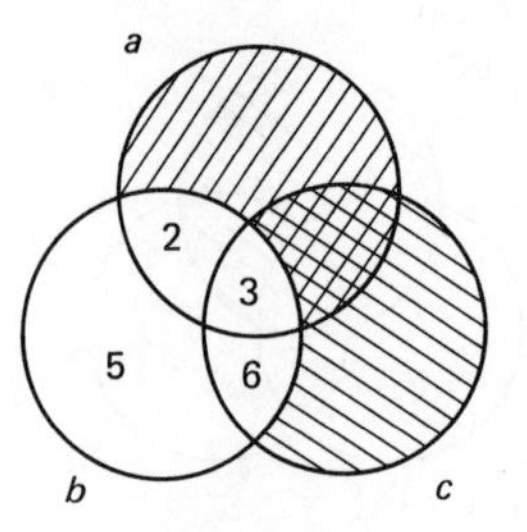

Figure B.24

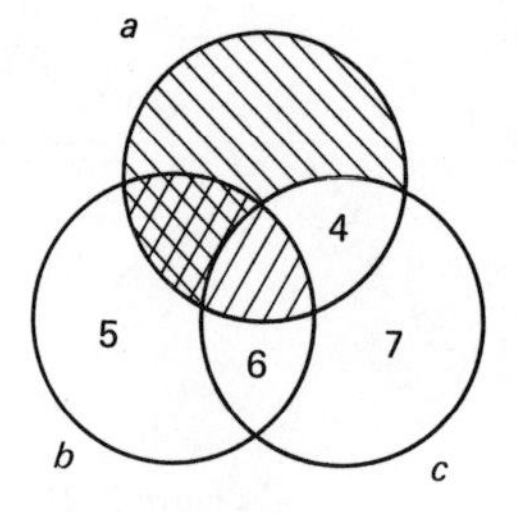

Figure B.25

2. All *a* are not *b*
   All *a* are *c*

   All *c* are *b*

Invalid (Figure B.25).

3. All *a* are *b*
   All *a* are *c*

   Some *c* are *b*

The conclusion says that some *c* are in the area 3 + 6 in Figure B.26. If we knew that there were any members of the class *a*, then area 3 would be occupied, and the conclusion would be valid. However, the premises don't shed any light on the existence or nonexistence of class *a* members. Consequently the argument is invalid.

4. All *a* are *b*
   Some *c* are *a*

   Some *c* are *b*

Valid (Figure B.27).

5. Some *a* are *b*
   Some *c* are *b*

   Some *c* are *a*

Invalid. The premises don't require that area 3 be occupied (Figure B.28).

6. No *a* are *b*
   Some *b* are *c*

   Some *c* are not *a*

Valid (Figure B.29).

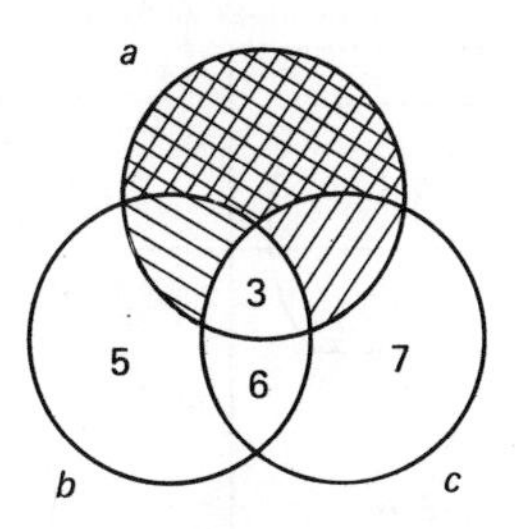

Figure B.26

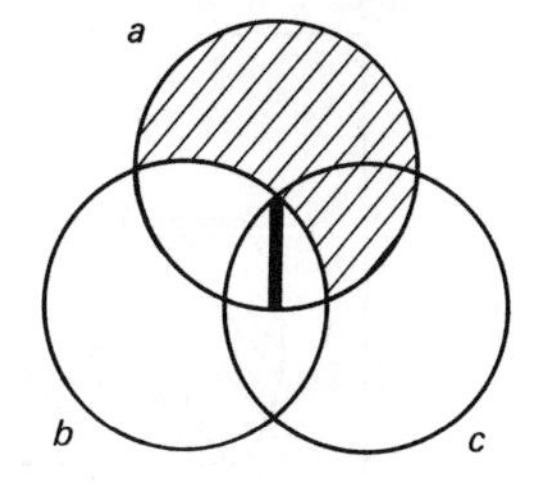

Figure B.27

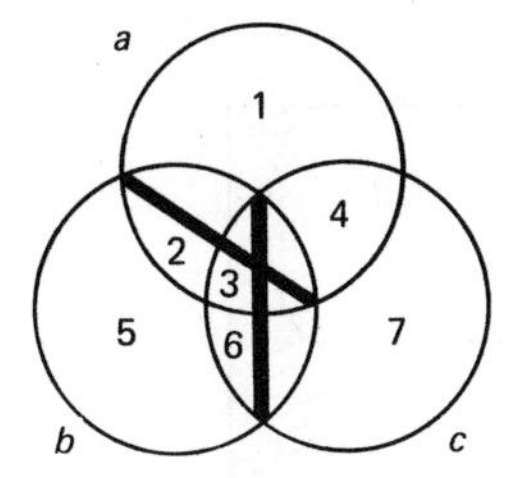

Figure B.28

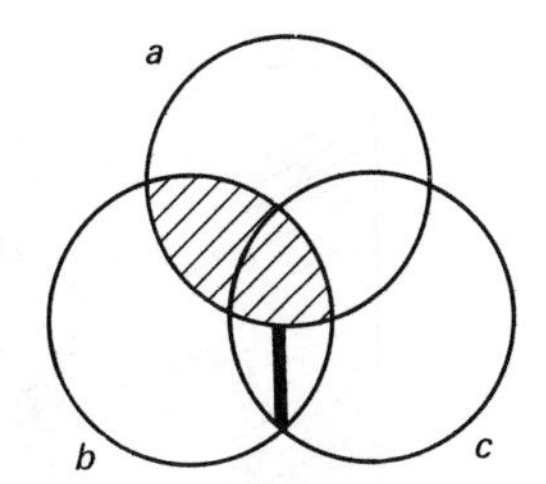

Figure B.29

B.16. INVALID ARGUMENTS

It's useful to know how to construct a valid argument. It's also helpful to be able to recognize invalid arguments. Some of the more common invalid arguments are as follows.

B.16.1. Asserting the Consequent

The form of this argument is:

$$\begin{array}{l} A \rightarrow B \\ \underline{B \qquad} \\ A \end{array}$$

Analysis of this argument is as follows. The implication $A \rightarrow B$ means that the proposition A & B' is false (Figure B.30). Now if B is true, we get the diagram shown in Figure B.31. The diagram indicates that both A and not-A are still possibilities. The argument is invalid.

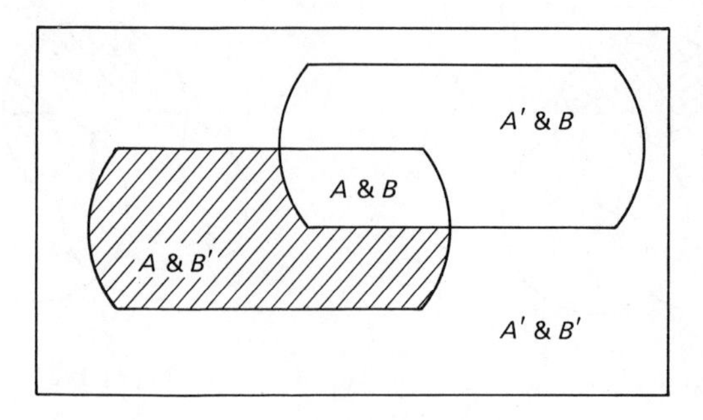

Figure B.30

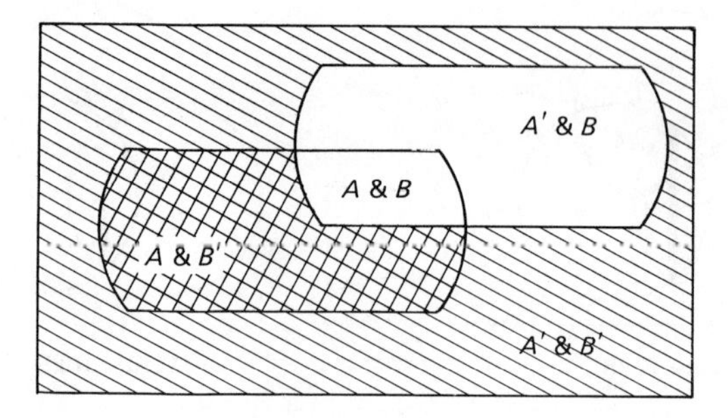

Figure B.31

B.16.2. Denying the Antecedent

The form of the argument is:

$$\begin{array}{l} A \rightarrow B \\ \underline{A'\quad\;\;} \\ B' \end{array}$$

The implication $A \rightarrow B$ means that A & B' is false. If not-A is also true, we have the diagram shown in Figure B.32. Both B and not-B are still possibilities. The argument is invalid.

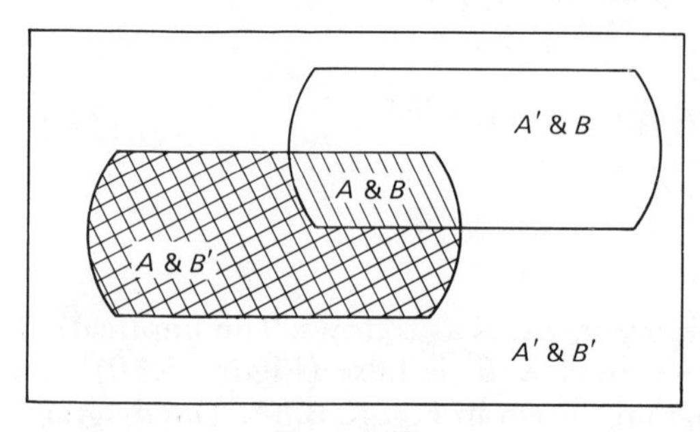

Figure B.32

B.16.3. Invalid "Chain" Arguments

The form of a chain argument is:

$$\frac{\begin{array}{c} A \to B \\ B \to C \end{array}}{A \to C}$$

An invalid argument form that looks similar to a chain argument is as follows.

$$\frac{\begin{array}{c} A \to B \\ C \to B \end{array}}{A \to C}$$

The implication $A \to B$ means that A & B' is false, the implication $C \to B$ that C & B' is false. Inspection of the diagram in Figure B.33 indicates that A & C' isn't necessarily false, so the implication $A \to C$ doesn't follow.

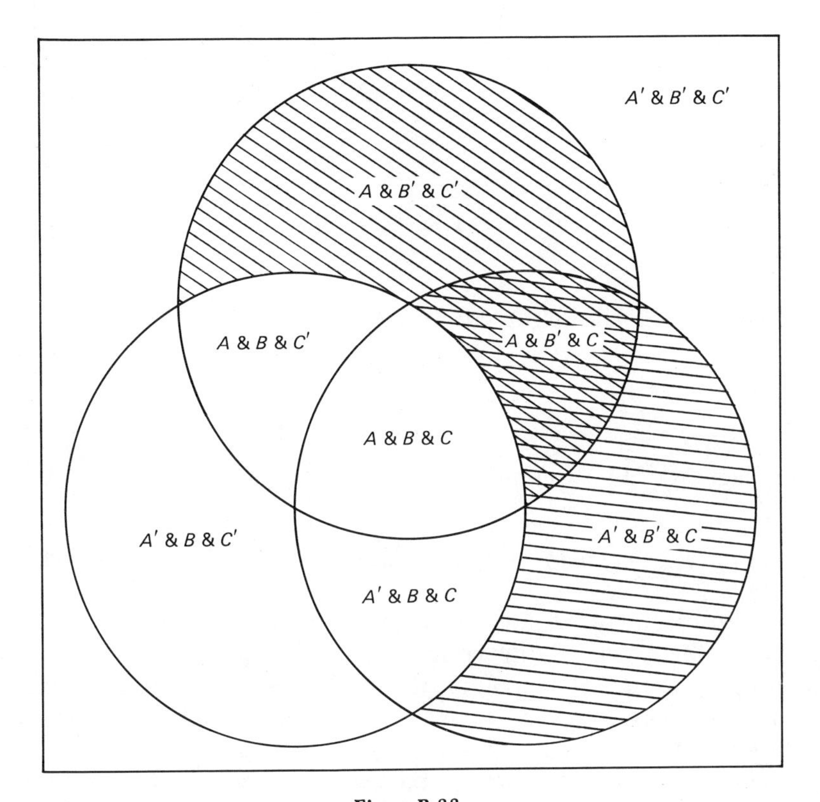

Figure B.33

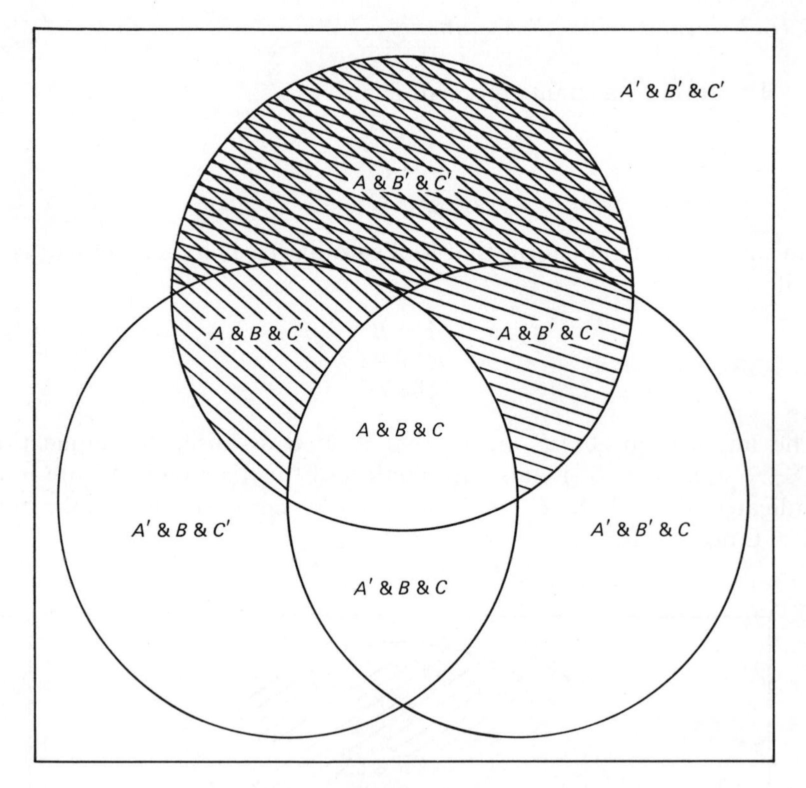

Figure B.34

Another invalid argument form that looks similar to a chain is:

$$\begin{array}{c} A \to B \\ \underline{A \to C} \\ B \to C \end{array}$$

The diagram of this argument is shown in Figure B.34.

B.17. VALIDITY AND TRUTH

Validity and truth are different things. Even when an argument is valid, the truth of the conclusion is guaranteed only when the premises are true. So if you're doubtful about the truth of a conclusion, don't confine your investigation to the validity of the argument—question the truth of the premises as well.

The alternative argument is particularly susceptible to refutation by questioning the truth of its premises. The author of the argument must show conclusively that the alternation *A* v *B* holds, or the argument fails to establish its conclusion. For example, if the actual situation is *A* v *B* v *C*, then the demonstration of not-*B* doesn't establish *A*, since *C* may as easily be the case.

B.18. HOW DEDUCTION IS USED

A knowledge of deduction is useful for:

1. Refuting invalid arguments
2. Proving propositions

However, deduction seldom leads the user to conclusions he didn't previously recognize. New ideas simply aren't developed this way. The more usual situation is to recognize, without benefit of deduction, that a proposition is true and to then develop a rigorous chain of deductive reasoning to prove the point. Deduction by itself just doesn't provide adequate direction to be a fruitful exercise. First you must have determined the conclusion you want to reach. Deduction will then help you either:

1. Figure out how to draw the conclusion validly from true premises or
2. Demonstrate that the proposition can't validly be maintained

Index

S

T

U

V

W

Y

Z